MY SEARCH FOR TRUTH

# VOLUME I

## JACK R. TRUETT SR.
*Christian avatar*

# MY SEARCH FOR TRUTH

World Ahead Press is a division of WND Books. The views and opinions expressed in this book are those of the author and do not necessarily reflect the official policy or position of WND Books.

Hardcover ISBN: 978-1-946918-14-7
eBook ISBN: 978-1-946918-15-4

*Printed in the United States of America*

# DEDICATION

*This book is not only dedicated to establishing a great civilization, but it is specifically dedicated to the advancement of science and to those that wish to know more about the Earth and our solar system. I have written it in an easy to read style for your enjoyment.*

# CONTENTS

# GRAPHICS AND SPECIAL DATA

# PREFACE

We struggle within the kingdom of man to know truth about our reality. However, mammon has entrenched powers in every aspect of our social/cultural system and thus we are born in sin. Therefore, the struggle to know truth is difficult. Although as man struggles in the trek towards understanding our reality and our reunification with God there are forces created by God that are working through Christ to assist man in his travails.

Thus, by God's love for mankind, his son, Christ the Lord, kindles new lights during periods of tribulation, and as man emerges from troubled periods, some of those new lights become brilliant and serve as beacons for mankind. The books titled *My Search for Truth* are brilliant beacons that will guide man to new horizons about our environment and therefore to higher spiritual development.

Yes, higher spiritual development is a goal that we must work to accomplish. We are not struggling in vain, for during this time in man's development the Holy Spirit is being bestowed (poured out) upon all the Souls upon this Earth. Due to this stimulation, great agitation and great cogitation has resulted; therefore, the evil individuals and those against Christ are stimulated to be more evil and those dedicated to Christ are stimulated to bring forth great works.

The books titled *My Search for Truth* have come forth from the "Spirit of Truth" and directly from "Higher Consciousness" where the mind of Christ rules. Thus, the true source of the knowledge and information that is given in the three books of which I have been honored to be the Christian avatar is Christ the Lord. Thus the great truths that have been brought forth through myself by Christ the Lord is for the general plan of God and his son, Christ the Lord, that being, to Glorify God by the spiritual development of humanity.

# Christ's people are inventing and discovering.

The tempo of spiritual development has steadily increased since Christ dematerialized his body and ascended into heaven. One visible result of this increased tempo is the fact that Christians have been busy controlling nature's forces and capitalizing on nature's forces, for nearly all the technological inventions since the ascension of Christ have been by Christians. The few technological inventions that have come forth through non-Christians have been brought forth by very disciplined people. The antichrist people and those that do not know Christ have brought forth virtually nothing for the development of humanity. Many of these people are parasitic in nature and are a drag on humanities development.

# There are many antichrist people.

The greatest numbers of people that comprise the antichrist group are the descendants of Abraham and Hagar by a son named Ishmael.

Ishmael and his descendants were and are born with varying degrees of psychosis. In the book of Genesis, Moses predicted the mental state of Ishmael to be psychotic and is proven by the descendants of Ishmael's lifestyle in the last two thousand years. Moses wrote, according to historical text in the Old Testament that was given by Moses about 1,300 years BC, that the ancient roots of Israeli and Ishmaelite people began with Abraham, as recorded in the book of Genesis. The origin of these people began with a very wealthy herdsman in the Mediterranean region, probably Iraq. His name was Abram (his name was changed by God to Abraham). His people were of a tribe known as Semites, and moved often to find the best grazing land for their herds. According to biblical text, Abraham's wife, Sarah, had a slave (also called a handmaiden or maid) named Hagar. Abraham fathered a son by Hagar that was named Ishmael. He was Abraham's oldest son; however, this son and his progeny were predicted by Moses to be a problem for all humanity. Centuries later, the Ishmaelite people were renamed the Arab people by the Greek historian Herodotus.

God revealed to Moses that an angel of the Lord had spoken to Hagar and revealed the nature of the child she would bear that was still in her womb. An account of this is given in Genesis 16: "And the angel of the LORD said unto her, Behold, thou art with child, and shalt bear a son, and shalt call his name Ishmael; because the LORD hath heard thy affliction. And he will be a wild man; his hand will be against every man, and every man's hand against him; and he shall dwell in the presence of all his brethren" (Gen. 16:11-12).

So the patriarch of the Ishmaelite (Arab) people was born with a severe mental aberration or psychosis. Ishmael was not the only progenitor of abnormal offspring, for there have been several impaired patriarchs recorded in the history of man, most notable the Jukes family - and the Neanderthals. The African pygmies should be added, since the original native that had an afflicted growth hormone brought forth his own kind. Although whoever he was, he passed this abnormality to his offspring and thus he became the patriarch of the African pygmies. Also going back four thousand or more years, in the land known as Mongolia, a Mongoloid baby was born that was believed to be a god, and the people that became known as Mongols had sexual relations with the Mongoloid and that person became the patriarch of the Mongols.

Abnormalities are not restricted to humankind for warm and cold-blooded creatures have brought forth abnormalities such as two-headed snakes, turtles, and other oddities, although these oddities have not created a new strain. However, this researcher believes the hammerhead shark - is from a strain of abnormals that goes back eons to a patriarch shark that had an abnormal gene configuration that caused the hammerhead shark to have the eyes on a protruding stalk.

# As the centuries passed, the genetic strain of Ishmaelites multiplied.

In the sixth-century BC, a Greek writer by the name of Herodotus discovered the Ishmaelite people living a nomadic life on the land now known as Arabia. He named these people Arabs since the word *Arab* in their modified Phoenician/Hittite language meant "wanderers or nomads," which corresponded to the Greek word - *nomades.*

The Ishmaelites had splinter tribes that lived a nomadic life on the land of Arabia until the year 570 AD, when a boy was born that grew into a tribal leader and was successful at combining the tribes into a cohesive unified tribe by establishing a philosophy that bore his name–Mohammed. Thus Muhammadanism became the name of their faith. Another name that was adopted for their faith was Islam, which meant "surrender to god." So from the beginning, Jews and Ishmaelites are half-brothers; however the offspring of Isaac and Ishmael were destined to be vastly different. Isaac's progeny became Israelites through prayer and blessings and the two half-brothers lived different lifestyles. The Ishmaelites appear to have been born with a very self-centered nature from the genes of Ishmael or this psychological aberrant became adopted with the adoption of Mohammed's directives. Whichever be the case, those that adopted Mohammed's philosophy became very self-centered. The psychological aberrance bound Ishmaelites to evil creeds that were given by Mohammed to his followers. Thus, the Ishmaelites believe that their philosophy has come from god and all others that are not of their faith should be murdered and god will reward them for killing the non-Islamists. The first psychotic aberration that was recorded as committed by the followers of Mohammed was to kill Jews, after which the directive was expanded to kill all that are not of the Mohammed faith. The psychotic directives ordered their members to subdue all lands where they immigrate to and establish their Mohammed faith.

Although there are many that have adopted the evil philosophies of Ishmael—that is, they are grafted in and are not genetically Ishmaelites—yet once adopting the evil philosophy that IS written in the Koran, these people are dedicated to a creed set forth in their philosophic tenets that are recorded in their book of directives from Mohammad. Many of these people live in Africa and the Middle East. These people are generally referred to as Muslims; they have not brought forth any technological inventions, yet are eager to acquire and use the devices and products that were invented and manufactured by Christians and Jews in order to improve their own socio/cultural structure and use the Christian/Jew technology in open warfare and terrorism against them.

This should not be a surprise to the learned and wise, for if they read the book of Genesis, where Moses prophesized over three thousand years ago: the decedents of Ishmael known as Muslims today would be afflicted with a destructive mania and would be a wild people or be uncivilized. In Genesis 16:11 and12, it is written: "And the angel of the LORD said unto her, Behold, thou art with child, and shalt bear a son, and shalt call his name Ishmael; because the LORD hath heard thy affliction. And he will be a wild man; his hand will be against every man, and every man's hand against him; and he shall dwell in the presence of all his brethren."

Thus the Ishmaelites and their descendants are and will be corrupted with a cursed mind that will extend to an abnormal group mind in heaven.

Therefore, the prophecy of Moses comes true in the verse that states "a corrupt tree cannot bring forth good fruit." The prophecy of Moses has come true and will continually come true until the descendants of Ishmael are nearly obliterated during the battle of Armageddon (Megiddo) in northern Israel. However, at the present the evidence that the tempo for mankind's spiritual development is increasing is abundant; therefore, I entreat all Christians and in a larger sense all of God's people that are willing to assume some moral responsibility, to work diligently in an effort to build great civilizations that are showcases with mentally healthy people where beauty, order, cleanliness, and great interrelated systems are a prelude to the coming "Kingdom of God upon this Earth."

# A NOTE ABOUT TRUETTS

The clear truth about the origin of these works is the fact that every person that aspires to serve God and Christ the Lord may become an instrument for their purposes and thus become an avatar. This is not difficult to understand, although the mind of man is very complex and when I state that I give thanks to God the father within me, my super conscious mind, my Father within, and Christ the Lord for revealing secrets of the universe to me, this can be better understood after the reader studies consciousness, in Volume III.

## All of us are born with challenges to meet and obstacles to overcome.

The general path of my destiny was set in my unconscious mind before I was born. As I review my life, the events that shaped my life has revealed a clear pattern that I was born to be a scientist. However obstacles and challenges began when my mother died, in 1930, at my paternal grandparents home in the same bed that I was born. At that time I was two years old and my sister was ten months old and six years later, in 1936, my father died of a heart attack when I was eight-years old. Thus my Pennsylvania Dutch grandparents became my parents. Years later after I had matured my grandparents and uncles told me that I was an extremely curious little boy that was always asking questions and others told me that I seemed to analyze everything. Due to these traits I became a deep research person. My searches have truly had a wide swath: due to this I cannot remember the many sources where I searched for knowledge for I have always been searching for knowledge and understanding.

A compelling urge to search for truth is evidenced from my avid interest in science and the nature of man's mind. Science magazines, encyclopedias, textbooks, and history of science books have been great places for me to concentrate my mind, away from worldly interests. Since Google has become such a wonderful search tool, I have used Google when other sources did not supply the information that I was seeking, even though I have an up-to-date encyclopedia installed in my computer.

The history of science and its renowned scientists have always enthralled me. I have read about the works of other renowned scientists that preceded me, with a strong focus on the lives and works of Copernicus, Kepler, whose work was made possible by the exacting recordings of planetary movements by Tycho Brahe, then Galileo. Einstein's theory of general relativity became a especially rewarding intellectual experience. While the works of other physicists and astronomers have enabled me to climb the mountain of understanding, it was the work of Kepler and Einstein that occupied more of my study time than any other scientist in the disciplines of physics and astronomy.

My interest in our solar system began when I was very young. I drew a picture of our solar system when I was in fifth grade and it hung on my bedroom wall for many years. The interest in the solar system eventually led me to study Einstein's theory of general relativity. It became a chief study when I learned that Einstein claimed that the planets revolved around the Sun by inertia. This postulate of Einstein's attracted me to delve deeply into the general working hypothesis of general relativity because I reasoned that all nine planets are known to revolve in the same counterclockwise direction as the Sun's counterclockwise rotation and this condition could not have been caused by accident or happenstance.

Also, the fact that all the planets are located in our solar system in a position that their distance from the Sun is mathematically cubed is equal to their distance squared positively could not have happened by happenstance. This evidence refuted Einstein's idea of the planets just being accidentally located where they are and revolve by inertia. A generally known law of psychology known as the law of determination" states that "nothing just happens, everything is caused." This law also applies to the physical sciences, so the planet's revolutions and their distance from the Sun could not be attributed to happenstance. Rather, by strong logic, these conditions were caused.

Thus these claims of Einstein stimulated me to became deeply attracted to studying general relativity, and I could not put it aside until I knew how and why Einstein developed his famous theory. I bought many books about Einstein and his theory of general relativity until I was able to take it apart and put it together like a jigsaw puzzle. It was difficult, because some postulates in Einstein's theory were extraneous to solar system dynamics and are not bound together in one cohesive paradigm. Yet I persevered, and finally I understood it clearly and I realized that general relativity was a false theory. Einstein's theory of general relativity did NOT offer a rational to explain the origin of the planets. His focus was on WHY the planets did not get attracted to the sun, since the Sun has such enormous gravity.

To support his idea, Einstein created an idea of troughs that held the planets in place and the idea inertia was responsible for the planet's revolving motion. This idea was a shallow concept that ignored an already established law of planetary dynamics set forth by Johannes Kepler. Kepler's third and true law states that a planet's distance from the Sun cubed equals its distance squared. And the rationale supporting this law is the fact that the Sun has the force and uses its power to drive the planets in their orbits with clockwork precision. So the proven facts of our solar system were superimposed by Einstein's opinions and fanned into believability by a writer of the *The Times* of London.

The facts in my review of history are true; however, please read on, for these bold claims in this work which you are reading without a doubt refute Einstein's theory of general relativity and place Kepler on top for contributing new laws that cannot be refuted. And his laws actually present a greater understanding about planetary motion.

Later in life i became enthralled with my studies about psychology, because if planetary motion affected the mind of man, it was a great challenge to know how. When I made an in-depth study of the mind, the works of Karl Gustav Jung, Freud, Adler, McDougall, and others, they became just as rewarding to my voracious inquisitiveness as the scientists that worked in physics and astronomy.

There is a law in psychology called the law of compensation. It is nature's way of bestowing talents and attributes to a child that was born with a birth defect or in early life had an accident causing a physical disability or a mind/body malfunction.

Examples of individuals that were changed due to their mind/body mechanism lacking a particular talent or attribute are Johannes Kepler, Charles Steinmetz, Alec Templeton, George Shearing, Art Tatum, Helen Keller, and Derek Paravicini. In addition, there have been many known savants in our history, and in our time several have been the stars of TV programs that focused on savants.

There are other talented individuals whose talent can be attributed to the law of compensation; however, the given examples are sufficient to illustrate people that are either an extreme genius in one or more abilities or are autistic and are gifted due to mind/body impairment with one specific ability. These gifted people have caused me to often wonder if my mind was changed due to a physical disability and early life trauma.

The fact that my mother died from influenza in the same bed in which I was born, when I was two-years old, and my 10-month-old sister and I were left in the care of our Pennsylvania Dutch Lutheran paternal grandparents, J. B. Truett and Jennie L. Gladfelter Truett, had a traumatic effect on our minds. When our mother died, we were supposed to stay at our grandparents' home until our father could take us to our home to 1040 Edison Street in York, PA. However, our father was psychologically devastated and we never returned to Edison Street. Six years after our mother died, our father died of a heart attack.

## THE GLATFELTER AND GLADFELTER LINEAGE

Sometime in the last several hundred years a registrar recorded a Glatfelter as a Gladfelter, and so began the beginning of two different spellings for the Glatfelter name, although there are other names that registrars can be credited to different spellings as Glotfelter, Clotfelter and other names. The lineage on all Glatfelters, Gladfelters, Glotfelters, Clotfelters and other misspelled names for Glatfelter are traced to a Swiss immigrant by the name of Casper Glattfelter, who landed on American soil about 1723. After moving several times, he eventually settled on land that resembled his home in Switzerland. Decades later, after the railroad cut through his land and the few residents in this farm community wanted a post office, and because their community needed an official name, they decided on Glatfelter's Station.

## More people of the Germanic lineage immigrated to America.

The Mennonites and Amish also emigrated from Switzerland, and these Swiss immigrants are all blood related. Therefore, I am blood related to the Germanic Mennonites and the Amish, and it is no surprise that I feel a cultural bond to the Mennonite's plain life style.

My lineage in the Glatfelter tree begins with Casper Glattfelter. Casper begat Felix, who begat Daniel, who was born 1786-1837/8. He was my great-great-great grandfather. He was the eighth child of Felix Glatfelter and a grandson of Casper. Since Casper Glattfelter died one year before the Revolutionary War, Daniel never saw his grandfather. At the age of 21, Daniel married Margaret Emig, whose immigrant parents came to America from Germany in 1743—the same year that Casper came from Switzerland.

Daniel Glatfelter and Margaret Emig begat George, who

begat Wesley, who begat my maternal grandmother Jennie L. Glatfelter. She married a Truett, and they begat my father, James Henry Truett. He married Irene Frances Mort, and they begat Jack Ronald Truett, and that is me!

My grandfather Wesley lived in a Germanic village (Pennsylvania Dutch) named Seven Valleys. It is between the small towns of York, New Salem, and Glen Rock. From an early age, Wesley was active in civic affairs. He was a school teacher and a justice of the peace in North Codorus Township, and in 1885 he was commissioned to serve as Recorder of Deeds for York County and served one term.

He married a Swiss Germanic descendent by the name of Melinda Catherine Roarbach, who lived near Seven Valleys on what is now named Roarbaugh Road. From this union they begat one son and six daughters; my maternal grandmother was born in 1878 and was the fifth born of seven.

# The Glatfelter lineage brought forth many educators, businessmen, philanthropists and political leaders.

Representative William F. Goodling (1927- ) was born of the Glatfelter lineage since his mother was a Glatfelter. He was elected congressman for the 19th Congressional District in 1974 and served until his retirement in 2001.

George Michael Leader (January 17, 1918–May 9, 2013) was born into the Glatfelter lineage since his mother was a Glatfelter. George Leader served in the State Senate for the 28th district for one term and was elected as governor in 1955 to 1959.

During a 2005 Glatfelter family reunion the three of us happened to be at the reunion at the same time and I had my picture taken with the two esteemed gentlemen.

Three related through the Glatfelter lineage       Ex Governor George Leader
picture taken at a Glatfelter reunion
                                        Jack R. Truett. Sr.
Ex Congressman
Bill Goodling

### TRUETT, TRUITT, AND TRUIT LINEAGE

My paternal great grandfathers' surname was Truett. His lineage has been traced to a Norwegian immigrating to England over a thousand years ago. The name has been changed many times; being adjusted for these people that moved to different political boundaries in England.

**A TRUITT MAKES HIS HOME IN AMERICA BEFORE 1650.**

The first Truitt (a George Truitt) came to this land about 1640 or 1642 and others came later. The first George Truitt is listed in *Virginia Immigrants Records* (Volume 5, State Land Office 20), in 1652. He initially settled in Northampton County, Virginia, but later moved to Accomack County. This Truitt endured persecution because of his religious beliefs since he was a Quaker, and like Mennonites and Brethrens, Quakers did not and do not have ordained ministers, are against slavery, and against debauchery, for their righteous claim was and is obey the Ten Commandments and the rules of "good mental health" to keep a healthy mind and body. About the time of this George Truitt's death in 1670, because of their persecution, some family members moved to Somerset County, Maryland and later others moved to southern Delaware, according to the researchers Martin-Grubbs.

**TRUETTS, TRUITTS, AND TRUITS HAVE BEEN COMING TO AMERICA FROM ENGLAND FOR CENTURIES.**

The surname has two chief different spellings in America, these being Truett and Truitt. A Truett had a family with his first wife, and after she died he remarried and the children of his second wife were named Truitt. This lineage of Truett and Truitt has been the anchor for many incarnating Souls that became researchers, writers, scientists, and theologians. According to one researcher, Isaac Newton (December 25, 1642–March 20, 1726/7) became the most famous scientist in the lineage of the Truetts. Isaac Newton was the first to submit the idea, in 1704, that the light wave had minute concentrations of energy that he named corpuscles. Newton's term for light particles as corpuscles lasted for a little over 200 years until a brilliant American physical/chemist by the name of Gilbert Newton Lewis (October 23, 1875–March 23, 1946), in 1926, renamed the particles from corpuscles to photons. Later, the physicist Arthur Holly Compton (September 10, 1892–March 15, 1962) discovered that light had a dual nature, having both a wave structure and a particle structure. In 1961, this metaphysician, Jack R. Truett Sr. (1928- ) renamed the light particles to ethons.

## In America the name Truett and Truitt is known to be active in the theological, military, and political history of our nation.

**A TRUITT WAS TO BECOME A GOVENOR AND WAS A RATIFIER OF THE US CONSTITUTION.**

The first Truitt to make significant contributions to growing America was a ratifier of the U.S. Constitution for the state of Delaware in 1787. George Truitt was this civic active gentleman's name. He served in the Delaware State House of Representatives from 1788 to 1793 and the Delaware State Senate from 1802 to 1807. He then served as Delaware's 18th governor from January 19, 1808 to January 15 1811. He was the 12th governor to be elected to that office. George Truitt died in 1818 at the age of 62, and internment was near Felton, Delaware. In 1903, the Delaware General Assembly voted to reinter Governor Truitt, his wife, and daughter, at Barrett's Chapel, Frederica, Delaware.

**ALFRED MARION TRUIT WAS A PROMINENT CONFEDERATE GENERAL.**

From my searches, the next prominent Truitt was a Texas Confederate Brigadier General by the name of Alfred Marion Truit. The first of this line of English Truits settled in North Carolina. The Truit family left North Carolina in the fall of 1838 and entered Texas on March 3, 1839, at Pendleton Ferry. Members of this family became very civic-minded individuals in Texas, to the extent that their involvements in civic affairs are recorded in Texas history in Shelby County, TX.

While General Truit was one Confederate during the American Civil War, there are 103 Truett Confederate Records. There are also 50 Union records of Truetts, and on the Pennsylvania monument at Gettysburg the name William Truett is cut in granite with others in memory to Pennsylvania's soldiers that died during the Battle of Gettysburg. So Truetts, Truitts, and Truits were on both sides of the Civil War.

**GEORGE WASHINGTON TRUETT**

George W. Truett was a famous Baptist minister in Texas. His deep-set eyes reveal a man of deep perception, and he used this faculty to relate biblical teachings with everyday life. The future famous theologian, the seventh of eight children; was born in a log cabin on May 6, 1867 in Hayesville, Clay County, North Carolina, and died July 7, 1944, at the age of 77, in Dallas, Texas. His accomplishments and contributions to helping others is so vast that I recommend all interested in Dr. George Truett's history go to Google and discover Dr. Truett's great history. George W. Truett became pastor of the First Baptist Church in Dallas in September 1897, where he served until his death in July 1944. An elementary school in Dallas bears his name, and Baylor University's Board of Trustees officially reserved with the Secretary of State of Texas the name George W. Truett Theological Seminary in the event the board decided sometime in the future to create a seminary. One year later the Theological Seminary was chartered.

..............................................................................

# A shared trait of the famous Truetts, Truitts, and Truits is a passion for their work or intensity in their activities. This trait is obvious in the five books (so far) that I have written.

..............................................................................

## QUINCY HIGHTOWER TRUETT

This outstanding Truett became famous during the (grossly wrong and illegal) Vietnam War for giving his life that others might live.

Quincy H. Truett was born April 3, 1932, in Quitman, Georgia. He enlisted in the Navy on September 14, 1951, and served on the destroyers *John W Thomason* (DD760) and *Floyd B. Parks* (DD884) during the Korean War. Subsequently, he served successfully in *Tulare* (AKA1 12), *Chilton* (APA 38), and *Tawasa* (ATF 92). From August 1965 to September 1966, Truett served with the Naval Support Activity, Danang, in South Vietnam. Following a tour of duty with the Key West Test and Evaluation Detachment, he trained at the Naval

Amphibious Base at Coronado, California, in preparation for his return to Vietnam. On September 13, 1968, Chief Boatswain's Mate Truett transferred to River Patrol Flotilla 5 and joined River Division 551.

On the night of January 20, 1969, his division was operating on the Kinh Dong Tien Canal, where Truett was patrol officer for two of four river patrol boats escorting an armored troop carrier. The entire unit came under intense enemy fire, and the boat ahead of Truett's erupted in flames. When the blaze forced the patrol boat's crew into the water, Chief Truett ordered his own craft to the rescue. Seemingly oblivious to the enemy gunfire streaming at him, Chief Truett maintained a constant covering fire throughout the rescue of the crew of PER 8137. Though mortally wounded, Chief Truett continued to fight until the successful completion of the rescue. Chief Truett died of his wounds that same day. For his ". . . extraordinary heroism. . ." Chief Boatswain's Mate Truett was posthumously awarded the Navy Cross. In addition, a Navy frigate, *USS Truett* was named to honor this brave chief boatswain's mate, Quincy Hightower Truett.

My own self-taught education and my insatiable urge for scientific research has already been given. My broad range of research places me in the category of a metaphysician. I am not the first metaphysician in history, for Carl Gustav Jung was a metaphysician; however, he never made this claim and he has never been awarded this honor. Therefore, I am the first metaphysician since Isaac Newton that has brought forth new knowledge and my travails should be recorded in the history of science since my peers have not only been disrespectful to me but have been cruel. I have been laughed at, lied to, and have endured the stupidity of PhD's contending with me using opinions against my facts. Great universities such as MIT, Princeton, Penn State, Stanford, Berkley, and others in America, plus leaders in prestigious universities in Europe have refused to review my findings.

USS Truett (FF-1096)

| Career (US) | |
|---|---|
| Ordered: | 25 August 1966 |
| Builder: | Avondale Shipyard, Westwego, Louisiana |
| Laid down: | 27 April 1972 |
| Launched: | 3 February 1973 |
| Acquired: | 24 May 1974 |
| Commissioned: | 1 June 1974 |
| Decommissioned: | 30 July 1994 |
| Struck: | 11 January 1995 |
| Motto: | *Dédication à dieu et patrie* Dedication to God and Fatherland |
| Nickname: | "Do it Truett" |
| Fate: | Leased to Royal Thai Navy in 1994 and eventually sold 9 December 1999 |

It has been the same story for decades; you claim that Einstein's theory of general relativity is false, you claim that astrology is a true science, you claim that energy dissolves—ahem—which is counter to the law of conservation of energy, you claim to know the nature of electrons, Mr. Truett you lack comprehension—don't call us, we will call you. My life has been a continual struggle to make ends meet with a bout of depression from continual rejections.

The efficient and pleasant homes that I provided for my family have been made possible by buying second-hand tools, borrowing tools, using some used building materials and my own efforts in building desks, cabinets, and other projects that changed my environment to be a place for my family to enjoy life.

As I edit these manuscripts again, I am approaching the age of 89 years and I need a miracle to enable the greatest work in the last two thousand years (maybe five thousand years) to be published and brought to the pinnacle where all science books will be made obsolete and mankind will be provided with the information to establish a new and better world.

## BEING BORN INTO A CERTAIN ENVIRONMENT AND A CERTAIN FAMILY IS NOT AN ACCIDENT.

My Soul was attracted to the Truett and Glatfelter lineage to enable a source for scientific acumen to emerge through the genetic programming, Although the path was not direct, for there have been many stumbling blocks through life and I have tried to make each a building stone. According to two psychic researchers, my Soul made several sojourns in ancient Egypt. During one incarnation, my incarnate self-taught astrology in the secret schools of Egypt, where wooden balls were used to illustrate planets and the Naronic Cycle that is presented in Volume II was used in the solar system studies. This suggests to me that heliocentric positions of the planets were used.

During another carnal experience, I was told that I was again an astrologer in Frisia or Friesia where great observations were involved. I have wondered if during that sojourn my Soul's identity may have been known as Tycho Brahe.

In a later sojourn, I was told that I was known as a great piano virtuoso and in another sojourn I was told that I had become a master of the violin.

**WE SOUJOURN IN HEAVEN IN DIFFERENT DIMENSIONS BEFORE WE INCARNATE ON EARTH.**

In heaven there are different dimensions (planes of different vibration rates) that house Souls according to their state of spiritual development, (*in my father's house there are many mansions*), a hierarchy of power rules in each dimension that resembles the structure of a large corporation with Christ at the head. You can compare it to foremen, supervisors, general foremen, department heads, and plant managers at different plants. While in New York in a high-rise apartment resides the CEO and the Chairman of the Board. Christ can be somewhat compared to the chairman of the board, although in Christ's case he has God the Father's awareness and that means—the "sittith on the right hand of God the Father" does not mean that God's hand will become sore from his son sitting on his hand for centuries, it means that his son, Christ the Lord, has the same awareness as the father and the powers of God the Father can be activated by his son, Christ the Lord. In heaven the law of recompense is honored to the T (as on Earth) and a balance is sought to make even the extremes experienced by each Soul. I have wondered if my previous incarnations may have been weighted by outstanding professional lives and the council decided that I had to experience strenuous and conflicting feelings because my Soul's last incarnation was in the South where I was born into a wealthy plantation family and sided vigorously with the Confederate States. I suspect a life of cruelty to all that were below me in social status with a disbelief in Christ and immorality. If I am correct on this suspicion, then I have found a reason for many of my life's disappointments. In addition, I may have discovered the reason for the concentration of darkness or Ramafak in my right tibia.

The wisdom of our almighty heavenly Father has provided a way out of the predicaments that were created by our wayward ways. Being incarnate in this environment places each Soul with great challenges and obstacles. In my case, losing my mother when I was two and my father when I was eight were two traumatic experiences. Although a physical disability due to an injury to my right tibia was a greater traumatic experience.

I was only five years old when I had a disabling experience that changed my life. I became ill with osteomyelitis in my right tibia. That terrible affliction deformed my right leg. For my entire life it prevented me from engaging in any strenuous sports. After I was incapacitated due to the osteomyelitis, my mind turned to intellectual pursuits. After I became an adult, I tried to analyze my youth that was buffeted by psychological forces that I did not understand; this did help me understand forces that affect consciousness.

As the testosterone began to flow, my energy level seemed to explode. Building model airplanes, early crystal radio sets, collecting rocks, working on merit badges and riding my bicycle on twenty-mile rides was not enough to quell the explosion of energy. I was drawn to take an active part in the Mammon environment. Unknown to me at that time, ID forces were influencing my mortal mind consciousness. I can honestly say at the age of about 12, I could have been selected to be the least among my peers to bring forth great revelations from scientific searches. I was bright; yet I was without a controlled, concentrated focus. My hyperactivity was not a trait that inclines a person to staid scientific searches. My behavior at this time was at best unruly, and I was headed toward becoming a juvenile delinquent.

I started smoking at the age of thirteen and was skipping school. Certainly this was noticed by school officials and they were concerned about my level of achievement and my record of skipping school. I was bored with the slow pace and shallow academia. My grandfather had his own business, was a Sunday school superintendent, Boy Scout leader, and he was becoming frustrated with my behavior. Unknown to me at that time, other church members were concerned since I had always excelled in Bible studies.

I was unaware of the discussions about me; however, I was in trouble for hooking school, and soon I was before a wise friend of certain church members and was removed from my grandparent's home and placed on a farm in western Pennsylvania at Oakdale that was operated by a Pittsburgh Presbyterian Church group with some help from a Pittsburgh Masonic Order, for orphan boys and boys from broken homes.

The home was founded January 29, 1901 in Pittsburgh by Reverend J. W. Cleland (whose demise occurred in 1922) although before the founder's demise, it was moved to Oakdale and named The Boy's Industrial Home of Western Pennsylvania. It closed in 1972 when operating funds diminished below the needs to operate and the farm buildings were demolished. This home for boys was not destined to continue as a Boy's Town, although it preceded Father Flannigan's Boy's Town by several decades.

The general complex had four homes for boys of different ages. Each farm had a head master and a wise farmer that served as a father. The main building in Oakdale was the administrative center, and in the 1940's it was headed by a minister by the name of Reverend McMunn. Above Reverend McMunn was a Presbyterian Reverend Love, in Pittsburgh.

I was assigned to a dairy farm home that had a woodworking shop and housed fifty boys. I stayed on the farm for about two years and had two of the most memorable years in my life. Gone from my life were the boys that were experiencing maturing problems and my scientific acumen was given an environment that set free my love for science and it blossomed.

After eighth grade we attended high school in Oakdale. I loved Oakdale High school, and my academic achievements placed me in the top one percent of the school. My final grade for ninth grade was 96 and a fifth. Heading the list in achievement was science and algebra, with a 98 percent in both. I loved physics with strong ardor in ninth grade and I dreamed of becoming a physicist. However, as my destiny was unfolding, I was never to go to college to become a physicist. As the decades passed, I discovered the law "as ye sow, so shall ye reap" was greatly influenced by forces that were not yet understood.

That is me, in 1943, age 14 on a farm in Oakdale Pa. with dreams of becoming a scientist.

At the age of 14, I was the least interested in becoming a farmer. My heart was longing to become a physicist and the Lord had a destiny for me to serve him, and I found the objects of my desires in Metaphysics. My life was a twisting, turning, up-and-down experience. But my discoveries that Jesus of Bethlehem had become the first Son of God and was indeed an intercessor for we humans to God the Father, my love for the Lord grew and my love for science kept me going as a distant light beckoned me to keep on the path. My love for the Lord and my incessant searches to know more about our heavenly Father's universe has resulted in discoveries that will change the world. That's me where I work in my office. It's been about 70 years from working on the farm. From that 14-year-old orphan farm boy to a scientist in several disciplines is an incredible achievement.

My discoveries were to be revealed to entire humanity.

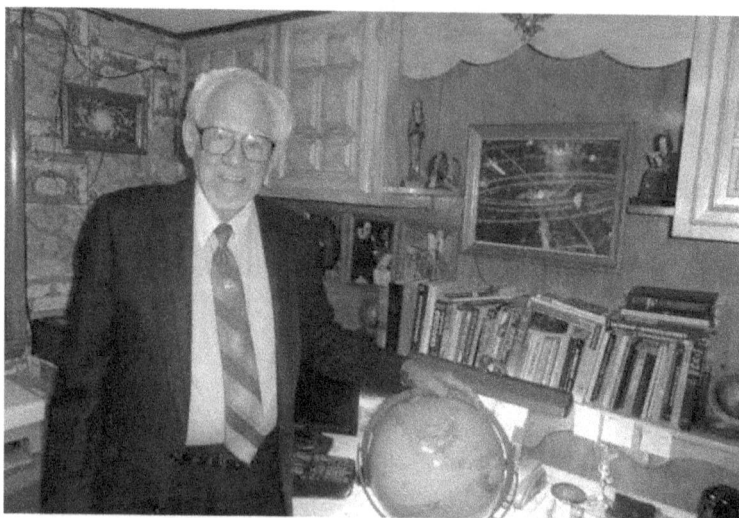

At the age of 16, in 1944, I returned to York from Oakdale. Then, when William Penn Senior High began in September 1944, I met a fifteen-year-old, five-foot two-inch tall, blue-eyed, blond-haired girl with pure German lineage named Betty Feltenberger. We were both humble orphans; my grandparents were very conservative Lutherans; and Betty's father was sired and born in a log cabin.

Although Betty and I had no inkling that September 1944 was to be a pivotal time in our lives as Betty and I began a steady relationship in September and it was to become a near 70-year journey together on this Earth, she was the first person that I ever met that could take my mind away from science.

**AN INDIFFERENT ACADEMIC ENVIRONMENT COMPARED TO OAKDALE WAS NOT FOR ME.**

I started senior high in 1944 with great expectations, although the hopes that I had to become a scientist were soon shattered, for I hated the school. Gone were my science buddies and teachers that took a part in student activities. I decided to go back to Oakdale, for I loved that school. However, this beautiful girl was filling my hungry Soul with love and understanding that I never before experienced in my life excepting when I was studying science. My emotional life was experiencing two attractions. I wanted to become a physicist and at the same time, I couldn't bear the thought of going back to Oakdale and leaving Betty.

Betty and I had talked about getting married and spending our lives together and the solution to my dilemma, or I thought at that time, was to quit school and prepare to marry Betty. A teenager getting married during World War II was difficult;

however, Betty and I had a relationship that was made in heaven and after several failed attempts to get married; due to the war years, in 1946, we finally got married in Baltimore, Maryland.

## My love for science was never quenched.

Even after we married, my mind was never detached from thinking about the reality in which we live, and after becoming parents four times and remodeling an old home that we bought, my mind returned to science. The same intensity that was concentrated on science before I knew Betty had again returned. Like a poet that seems unable to keep his mind away from poetry and the artist that lives an almost reclusive life to satisfy his urge to paint, my mind was always cross-examining scientific ideas.

In 1958, I discovered the astrological Naronic Cycle. I explain the nature of this chart in volume II. Before I discovered the Naronic Cycle, I had studied Einstein's general theory of relativity and I perceived a flaw that led me to sincerely doubt the validity of his theory. Somehow, this stimulated me to return to studying physics and astronomy like a voracious hungry wolf devours his prey.

After I had discovered that certain held beliefs in physics and astronomy were opinions and were never proven facts, I become frustrated from the fact that an amateur scientist could not bring forth a work that would be honored. Later in life, due to the heavy feelings of failure, I had several bouts with depression. Only by going to my work shop and building carpentry projects that I had conceived, creating several beautiful gardens and going to work as a delivery person for Advance Auto Parts on Mount Rose Avenue in York, and praying to the Lord when I got those depression feelings did my mind become refocused and those weird feelings from depression reoccurred less frequently and finally did not return.

All the scientists that submitted great works in their lives graduated from school, went to a school of higher learning, got a degree and a PhD, and taught in prestigious universities. Only a few that did not follow this path brought forth great works including, Alexander Graham Bell, Thomas Edison, and Bill Lear of the Lear Jet Company that he founded. These men were chiefly self-taught. There were others that dropped out of college and founded their own companies, such as Bill Gates and Steve Jobs, whose college education ended after eight months. Many musical geniuses never graduated from high school. Three of the most gifted composers never graduated, including Irving Berlin, George Gershwin, and Errol Garner.

## I was not in control of the thoughts and actions that charted my destiny.

My formal education ended with the completion of my ninth year. It was a decision that I would not advise anyone to take. The school of hard knocks is especially hard. At times during life I was asked what college I graduated from. I answered several times Howard Knox.

**A NEW LIFE BEGAN**
My married life started with plenty of love and plenty of nothing.

I bought a used bike to peddle my way to work, and bought some used clothing for myself. Within the German genealogical lineage of my people, there were educators, businessmen, and a surgeon. They all adhered to the philosophical principle "you sleep in the bed that you make for yourself." I later discovered that blessings from struggles usually multiplied and that which is given usually comes to naught.

Changes to my life began that were beyond my expectations and some beyond my comprehension. One year after Betty and I were married, I began to have psychic experiences. At that time I could not find books that dealt with the dreams and the out-of-body experiences that I was having. From these psychic experiences, my scientific curiosity was stimulated to learn about the realm of the metaphysical. I searched and I searched and eventually I found books that were classed as the occult. I read many books, and about 1958 I began to have conscious out-of-body experiences. In 1969, I went on a fast for 24 days, and this traumatized my system in a manner that caused psychic changes to occur with my consciousness. The affects of these two experiences had most remarkable changes to my life that led to greater changes as the years passed. It was about this time that my interest in physics and astronomy increased.

### GENIUSES DEPARTED FROM EUROPE AND CAME TO AMERICA

Dr. Leo Szilard and Dr. Edward Teller had departed from Europe and came to Princeton University. Albert Einstein had left Germany and came to Princeton University. Enrico Fermi left Italy and came to America. While these men were either teaching or working on theoretical physics, I was studying either parapsychology or metaphysics. Decades passed and the great men that I thought most highly of had departed from this Earth for regions unknown.

I have gone beyond traditional physics, meteorology, astronomy, anthropology, and psychology and I have become a metaphysician. I never belonged to social organizations, and I have lived a very private life. I know my success in finding secrets of the universe where others failed is attributed to the power of the Holy Spirit due to my faith in God and Christ. I believe my out-of-body experiences created a mini-programming that enabled me to become unified with my Soul mind or my Super Conscious mind during periods of concentration for the intricate and complex solar dynamics related to planetary orbs could never have been solved by a mortal mind.

It is now time for your intellectual journey to begin. Decades of research and years to write the three volumes are now yours for your education and amazement, and for the advancement of mankind. As these books are published and circulated, in time, great interest will be stimulated about my claims and revelations. Great discussions will ensue and a beginning of a new era will have begun. Christian nations will be the beneficiaries of new methods to control our environment. Volcanoes and hurricanes will be controlled, new sources of energy will make gasoline nearly obsolete and electricity will be plentiful.

I know that you are eager to discover *HOW* these changes are to occur, so I entreat all my readers to slowly and thoughtfully read all the volumes of "My Search for Truth."

# CHAPTER I

# PHYSICS AND ASTRONOMY ARE REBORN

The new science of the twenty-first century is presented in three volumes of *My Search For Truth*. These are: *New Physics, New Astronomy* (which you are now reading); *Music Of The Spheres: Forces That Affect Consciousness,* and a fourth book titled *Deep Truth*, which chiefly deals with political science.

*Music of the Spheres* is named in honor the great Pythagoras (about 582 BC–497 BC). and Johannes Kepler (1571–1630.) Between these two historic men, about two thousand two hundred years passed due to entrenched opinions about our solar system. But truth will eventually emerge from a morass of opinions by researchers determined to find truth among unproven opinions.

Now I state to you the reader and to all inquisitive minds that have been attracted to this work: You are about to enter into one of the rarest intellectual journeys within the entire history of humanity.

In Volumes II and III, great truths are presented in an uncomplicated manner about:

1. The nature of your mind.

2. Why humans behave as they do.

3. Why people are so vastly different.

4. How planetary interaction affects human behavior.

5. How astrology works.

This volume is specifically devoted to:

1. Energy and ethons of the ether and how energy dissolves.

2. The nature of the electron and electricity.

3. Our solar system's gyroscopic gravity field.

4. The nature of earthquakes, volcanoes, hurricanes, cyclones, and dust devils is carefully explained.

In addition to explaining how to control volcanoes, the terrible destruction by the hurricane named Sandy has prompted me to explain the method to control hurricanes; it is submitted at the close of this work. I state emphatically: hurricanes striking the eastern coast and the Gulf of Mexico can be ended and billions upon billions of dollars of property damage, human misery and loss of life will be prevented forever.

Typhoons striking Asian nations, such as Typhoon Haiyan that struck the Philippines on November 18, 2013 and caused 10,000 deaths and destroyed coastal areas, wreaking billions of dollars in damage will be ended forever. Monster typhoons and hurricanes like Haiyan cannot be destroyed or dissolved, but they can be steered away from land.

Here is a good side note. Muslims nations like Bangladesh that experience great destruction and loss of life when a typhoon strikes their nation can be spared. They won't be spared by Mohammed or by an imam that orders Muslims to kill Christians and Jews; they can be spared great destruction by another Christian scientist among those that brought Muslims into a life where amazing mechanical and electronic devices that make life so easy and amazing for them. A great side note, isn't it?

So, readers, put on the proverbial thinking cap and get ready for the great journey.


**EINSTEIN PROVEN WRONG**

Perhaps you were drawn to this work because you read or heard that a new understanding about our solar system has been revealed that proves astrology. If this is your chief interest, you will not be disappointed; however, the answers to your questions about astrology are not found in this volume, for the answer to all your questions about astrology can be found in Volume II, titled *Music of the Spheres*. This volume reveals a new dynamic of the solar system that enables or creates astrological orbs; it is the interaction of planetary orbs that affect the thinking and behavior of people.

For over 12,000 years, moving planets in the sky have been studied, and the belief in astrology began before 10,000 BC. That is a long time ago, yet it is highly possible that tens of thousands of years ago, early man looked at the starry heavens and wondered if those lights in the sky had any influence on his life. As time passed, man continued to develop, and as man developed, astrological research expanded, yet astrology has never been proven because the voluminous data needed was not available.

When the computer proliferated and the Internet was established, the United States Naval Observatory published an ephemeris that was accurate, and data from the NYSE became available. By using the data of these in correlation, I was able to do research that no other scientist before me was able to do. Although I knew that astrology was a true science and I fully belied in a God of the universe, or rather the omniverses, and I fully believed in God's son, Christ the Lord. Without believing in Christ and the higher consciousness available through Christ, I doubt that the secrets if the solar system would have been discovered, because it is too complicated.

Omniverse Defined: This word has reference to two environments, a physical and an etheric. Surrounding each planet, satellite, and star is an etheric environment. The Earth's etheric Omniverse is called heaven.

The new concepts of our solar system introduced in these books needed new words to explain new conditions and concepts and thus new words are introduced in each volume. *Aust*, *wake*, and *ethons* are new words that are introduced to physics and astronomy. Understanding the Aust and the Wake is essential for all students of physics and astronomy.

An Aust is a rotary gravitational force field that extends from all rotating celestial bodies at their greatest girth or zero line of latitude.

The Wake is an intensely perturbed area of gravity force above and below the zero line of latitude of each rotating celestial body.

With our Sun, the Wake's most powerful force is located about three and a half degrees on each side of the zero line of latitude. It sandwiches the Aust so the Aust's width is about seven degrees in width. The Wake begins a clockwise and a counter clockwise gravity field above and below the zero line of latitude.

The Sun's Aust is the force field that creates tubular confines that holds all the planets in their orbits and enables each planet to create astrological orbs. The tubular confines can be compared to the funnel of a tornado. As each planet revolves in its own confine, it moves to the upper and lower and outer and inner limits of its own tubular confine.

Before astrology can begin to be understood, it is necessary to understand the structure of the Sun's rotary, gravitational gyroscopic extended force field (about $7^0$ wide) that I have named the Aust. The nature of the Aust is carefully explained in this volume.

After you understand the Aust, Volume II explains how the orbs are created within the Aust and how their interactions affect human behavior. If you are interested in discovering how the laws of psychology and Christianity apply to a new science that proves certain planetary configurations affect human behavior, you will find answers to all your questions within the three volumes.

If you are interested in the mind and want to know what is the Soul, where is the Soul, what is the ego, where is the ego, what are phylogenic traits and other questions about the mind, you will find the answers to these questions in Volume III.

If your interest has been stimulated by the claim that Einstein's theory of general relativity has been proven to be a false theory, or that foundation ideas that make the framework of present physics and astronomy are proven to be false, you will find the answers to these questions in this volume, *New*

*Physics—New Astronomy.*

If your interest is more inclined to political science, in Volume III, you will discover that American wealth has been stolen by self-styled "Caesars" operating as Fabian Idealistic Dreamers. Contrary to Fabian Idealism, and more in line with dreamers of the Trilateral Commission and the Bilderberg Club, Volume III reveals the greatest unconstitutional act ever committed against the American people. This 1912, 1913 unconstitutional legislation, known as the Federal Reserve Act hijacked from the American people the constitutionally granted right to free coinage. The translation to modern-day usage is free issuance of legal tender.

Instead of honoring the right of free coinage, and creating federal banks for check clearing houses serving their own regional banks, the law was written to enable billionaires to skim a profit from every new dollar created by two schemes. These are creating bonds that charge interest to the American people for every new dollar created, and by charging a fee by illegally created bond salesmen for every bond created. This scheme and scam is carefully explained in Volume III. In addition, the destructive effects of following the false theories of the so-called gurus, John Maynard Keynes and Milton Friedman, is revealed.

In addition, you may have read or heard that destructive volcanoes can be prevented and want to know how these terrible eruptions can be prevented. In either case, a true and greater understanding about our environment is waiting for you.

Do not be astonished about these promises; be ready to learn; for you will discover that much that you have been taught about our reality has only been opinions and concocted tales to deceive you. Some of the deception is not much better than the myths of ancient Greece and Rome. So, be ready to embrace the truth presented in these three volumes, for this book which you are now reading is based upon new discoveries and will bring about the greatest changes in the affairs of man than any other single work in recorded history, excepting the Bible.

You will find how deceivers and plain liars have been working against the interest of American people to establish a welfare state and to favor the concepts of a few. You will find how special-interest groups have gained power in the classroom, government, and the business community to alter American destiny.

These changes were made in order to comply to serve political special interest groups and to serve education czar's narrow-minded false concepts of the universe in which we dwell, i.e., God does not exist, planets revolve by inertia, energy is always conserved and cannot be destroyed.

You will discover how the intent written into the Constitution—that is, to "promote the general welfare" and to "secure the blessings of liberty to ourselves and our posterity" has been ignored by using a demonic plan to give away American wealth and make America a police state.

For every curious individual that is seeking truth about our physical environment, you have been attracted to this work. Either you are a scholar of science or you have a desire to know truth about reality or otherwise you would not have been drawn to study this work. Your interest in peering beyond the veil that separates truth from false ideas and opinions will enable you to scrutinize abstract scientific principles. In addition, you will be able to understand how American politicians have created an enormous load of taxes upon the working middle class in order to serve their own craving to act as modern-day Caesars.

As a scholar of science, your comprehension of this work will be a vastly rewarding experience. If you have a desire to know more about the nature of our solar system, electricity and Einstein's theory of general relativity and have questions concerning these, you will find plainly explained answers to your questions in this volume that you are now reading.

There are great changes coming and I know that you want to know about them. The tempo of our advanced technical environment is increasing that enables changes to occur quickly. Read these chapters or volumes carefully so you can take an active part in helping to shape our days and years to come. Within the pages of this volume, evidence is presented that refutes presently held ideas and opinions that are the foundation of physics and astronomy. In situations where evidence is not available, tests are described that must be conducted to supply evidence.

**THE SCIENCES OF PHYSICS AND ASTRONOMY WILL BE REBUILT AND THIS WILL OPEN THE TWENTY-FIRST CENTURY TO SOLID TRUTHS, FALSE IDEAS WILL BE GONE FOREVER.**

The presently held ideas that will be refuted in this work are the claims of J. R. Mayer, Hermann Helmholtz, and Alfred Einstein. These false ideas are enumerated. The law of conservation of energy, that is—energy cannot be created or destroyed—is proven to be a conjecture that is 50 percent false.

The postulate of Einstein that attributes planetary revolving motion to inertia is proven to be false. You will learn what causes the planets of our solar system to be held in the ecliptic plane and why the planets revolve around the Sun; the idea that planetary orbits are ellipses I have also proven to be a false idea. Planets do revolve in an elliptical pattern within a *tubular confine* that has an inner and outer boundary. *The tubular confine can be thought of as a tornado's funnel that circles the Sun as a belt and holds a planet in its confine.*

Einstein's idea that ejected electrons from an electron emissive material when light is directed to the surface, comes from "ejected atomic electrons" is proven to be a false idea. If an electron emissive material, matter would eject its own electrons, it would then become electrostatically charged - and that has never happened.

Einstein's conjecture that "perhaps the electron has an inner action" is presented as *not a perhaps*, for electrons are proven to have an inner action. If the leading theoreticians in Einstein's time who were speculating about the nature of the electron would have studied a transformer, they would have realized that electrons are either formed or dissolved in a step-up or step-down transformer. This would have proven that, indeed, electrons do have an inner action.

The idea of Einstein's warped and deformed space that pertains to magnetism is proven to be a false idea. While Einstein did not understand magnetism, he steadfastly held to his warped and deformed space idea and from this idea he conjectured that Earth's magnetic field was a remaining mystery for scientists to solve. It was a mystery to Einstein as he declared, but a picture is worth a thousand words, and in this volume you will see the Earth's magnetism revealed in a drawing of the Earth's magnetic field.

You will find repetition in these volumes; however, please allow it to help you reinforce the knowledge that is presented. Understand, this volume is an introduction to how and why planets affect human behavior. It is necessary to grasp the true solar system mechanics referred to as the dynamics of the solar system. I want each reader to know how and why presently held ideas were adopted into science. Allow me to give a quick review concerning several critical aspects of the solar system that confounded great minds.

Einstein and other theoretical scientists were not completely satisfied with Newton's law of universal attraction. They knew it worked and therefore did not deny it, but something was missing in Newton's work. The probing theoreticians could not understand why the terrestrials were not pulled into the Sun, since the Sun contains over 98 percent of the mass of the solar system. Even if Venus, Earth, and Mars somehow could remain in a stable orbit, then why, oh why, could the little rock Mercury remain in orbit and not be pulled into the Sun?

**MERCURY'S PRESENCE HAS BEEN THE CAUSE FOR GREAT CONSTERNATION.**

As far back as records are available, we know that a small percentage of unusually gifted scientists have devoted much time into probing the mysteries of our solar system. The Mercury dilemma was one of these mysteries and was speculated about as far back as Kepler. Einstein was aware of the fact that Newton's Law was woefully deficient in explaining the stable orbit of Mercury and was eager to fathom the reason or reasons.

At some time in Einstein's theoretical thinking, he arrived at a seemingly reasonable possibility that some type of a trough in space provided a track and frictionless path for planets to revolve around the Sun. Einstein was convinced that if it were true, according to his reasoning, it would finally solve the riddle of Mercury's stable orbit. However, while Einstein was developing a theory to explain the unexplained, he drew upon the works of other scientists who had already theorized about the nature of our physical universe. A very good history of Einstein from the beginning of his school career to his demise and the scientists whose work Einstein copied can be found on Google. It is a must-read for all serious students of physics and astronomy. Although, the works of Coriolis, J. R. Mayer, Hermann Helmholtz, Maxwell, Lorentz, Max Planck, plus Michelson and Morley, were the chief and powerful influences in Einstein's theoretical reasoning.

Coriolis, Maxwell and Planck had submitted mathematical formulas to account for certain phenomena and Lorentz was known to be a super genius mathematician. Einstein studied their concepts with that of Buridan and probably Bruno and incorporated them with his idea of planetary troughs. Thus, a grand theory to be named the general theory of relativity was emerging from Einstein's study of the theoretical works of previous scientists. Einstein believed if he knit together concepts already accepted as gospel truth by physicists and astronomers about light, energy, motion plus gravity and wove them together in a complex theory with his planetary trough idea, It would not only answer the thorny question, what held Mercury and the other terrestrials in stable orbits, but would enable him to expand the scope of investigation to include the terminal velocity of light, relative motion in the universe (for which Bruno was burned at the stake for stating), and in his greatest theoretical gamble, to theoretically calculate the amount of energy in mass.

In my own evaluation, Einstein's idea of a kind of trough that served as a confine that held the planets in perpetual or locked position as they orbited the Sun was his own idea. I have not found this idea to be put forth by any scientist before Einstein. In truth then, Einstein's planetary trough idea was not only original, it was novel. Therefore, assuming the novel idea that planets were in a trough, the question remained, what caused these troughs? In order to submit a rationale for planetary troughs, Einstein concocted a false and very weird explanation that space warps and deforms in the vicinity of matter.

If this idea were true, the logical challenge to this idea would be: if space would warp and deform in the vicinity of matter, then why would space warp and deform? Einstein did not offer a rationale to support his idea, for he was stuck with his trough idea. It is logical to me that Einstein had created a closed theory that did not permit an alternate possibility.

Einstein knew his weak points, and one necessary trait for peering beyond present knowledge is perception that operates with imagination. Einstein confessed his weakness in an essay he wrote about his plans for his future while at Aarau. He wrote the following:

> "If I were to have the good fortune to pass my examinations, I would go to Zurich. I would stay there for four years in order to study mathematics and physics. I imagine myself becoming a teacher in those branches of the natural sciences, choosing the theoretical part of them. Here are the reasons which lead me to this plan. Above all, it is my disposition for abstract and mathematical thought, and my lack of imagination and practical ability."

That statement, "my lack of imagination and practical ability" lays bare Einstein's lack of perception and innovation. The lack of practical ability which can be paraphrased to be intuitive mechanical ability with his lack of imagination creates a great hindrance or completely rules out inventive exploratory thoughts.

However, Einstein's explanation for planetary troughs was a beginning of great truths but was underdeveloped. Einstein's idea of planetary troughs was not supported by any facts, except of course, since Einstein said it; we believe it. This is why he never received a Nobel Prize for his complex theory of general relativity, although Einstein was awarded a Nobel Prize in physics in 1921 for his *incorrect opinion* about metals emitting electrons when light is focused on the surface of the metal. This was referred to as the "photo electric affect." However, the emitted or ejected electrons were not and do not come from the atomic electrons of the metal, for some light is absorbed by the metal and are transformed into electrons—then ejected. 1. This idea of Einstein's had no relationship to his idea of gravitational troughs as proposed in his theory of general relativity, and 2. the planets revolve about the Sun by inertia.

I perceived planetary confines being created by a rotating spherical body without even considering Einstein's "gravitational trough" idea nor his assertion that planets revolved by inertia. I considered the fact that all the planets revolve counterclockwise just as the Sun rotates in a counterclockwise direction to be vital factors in the solar dynamics. I perceived Einstein's concept of solar dynamics to be greatly flawed and only compared this aspect of his theory with mine after I had developed a rational to explain the planetary confines and orbs. Later, I discovered that Johannes Kepler had discovered laws that proved why and how planets revolved about the Sun.

However, with regard to the age-old problem as to why Mercury and the terrestrials were not pulled into the Sun, Einstein was correct when he perceived as others perceived— Newton's explanation for planetary movements was lacking an explanation as to why little Mercury was not pulled into the

Sun billions of years ago.

Something in Einstein's idea of troughs seemed reasonable to him; however, his ignoring Kepler's third law and his rejection of Newtonian solar dynamics led to another opening and a further explanation as to why the troughs existed. Einstein rejected Newtonian solar dynamics and his warped and deformed space in the vicinity of matter filled that void. That idea seemed reasonable to Einstein, and he mistakenly incorporated that false idea into a new concept of solar dynamics.

Therefore, since he ignored Kepler's third law and rejected Newton's solar dynamics, he created an Einstein solar dynamics where inertia was the reason why the planets revolved around the Sun. The false idea of attributing inertia as the reason why the planets revolved around the Sun is preposterous and it created an eventual doom for his general relativity theory.

Great German and Dutch scientists were impressed by Einstein's explanation that accounted for the two closest inferior planets not being pulled into the Sun, plus the fact that great ideas from outstanding physicists as Lorentz were embedded in his theory. In this manner, Einstein was able to convince the greatest theoretical scientists to believe his theory of general relativity. However, not all were convinced. Yet, for almost a hundred years, many of the most brilliant physicists in the world were duped into believing Einstein's theory of general relativity. In addition, whether it is shameful or hilarious, depending on your point of view, PhD's have been granted on the basis of a thesis that was written about Einstein's false theory.

In this volume, I go into detail to explain how our present beliefs originated and why some of them are false opinions.

**SOME FALSE IDEAS WERE AS IMMOVABLE AS A GRANITE MOUNTAIN.**

Change has not always occurred quickly. Egypt existed for several thousand years. During that time, mathematics advanced, including practical uses for contrivances based upon mechanics. However, in the last four centuries, more changes have occurred than in the entire history of humanity. The pace of development was moved into a higher order when Michael Faraday (1791 –1867) invented the electric motor and the transformer. After Faraday invented the transformer over 200 years ago, the secret about the nature of the electron has been before science. Tens of thousands of scientists and science buffs have missed the fact that it is the building and collapsing field that dissolves and reformulates electrons in a transformer.

Following Faraday, Thomas A. Edison (1847 –1931) became the founder of the technological age. Although, it was Nikola Tesla (1856 –1943) who spearheaded the use of alternating current. Tesla eventually teamed up with George Westinghouse (1846 –1914), and it was the Niagara Falls Power Company that had installed Tesla-Westinghouse alternators in their generating turbines, and electricity became available for general use. The city of Buffalo, New York, was lit up with the use of alternating current; it was a victory for A/C current and a feat that achieved something that direct current could not have achieved.

Each new century has contributed to making a new world, and the new world is based upon technology. Our development pace will continue if we make some important changes. The power of the people can be a force that will bring forth great changes. From these changes that can come by demanding to know truth, a new and better world can be created. Then unfair systems, plus old beliefs and ideas will be wiped away in the early part of the twenty-first century, just as they have in the last four centuries.

It is logical that many of you will wonder how science in the 21st century will reshape our lives. Within the realms of political science (government), economics, and organized science, great changes must occur if we the people are to enjoy greater fruits of our own labors.

Our life styles can be reshaped by instituting changes in American government and demanding to know truth about politician's deceitful lies, cover-ups, and serving special interest groups against the interests of the vast majority. Within organized science, a spirit of honesty must replace the spirit of dishonesty. The idea of being politically correct instead of being honest must end and teaching fallacious theories in order to defend antiquated illogical ideas just to support personalities must end.

The demand for truth will bring into focus new vistas that otherwise would remain obscure. By this resolve, a new structure of the universe will open and the present fallacious concepts will be replaced by an expanded understanding.

### WE ARE NOW IN THE TWENTY-FIRST CENTURY.

As the first quarter if the twenty-first century commences, my volume *Music of the Spheres*, to use a metaphor, will knock scientists out of their shoes, for astrology is proven to be a true phenomena. The volume *Forces that Affect Consciousness* reveals a quagmire of problems that has been created by American politicians. By correcting our political problems, the consciousness of our minds will be enabled to rise to a higher plateau.

**MANKIND CAN RISE TO A HIGHER PLATEAU OF UNDERSTANDING AND A HIGHER STANDARD OF LIVING.**

From the truth revealed in Volume IV, *Deep Truth*, gasoline and coal can be replaced by five different energy sources. Hydrogen will become the staple energy that will be used to generate electricity, heat buildings, and power all internal combustion engines. All first-, second-, and third-class cities should be mandated to have built "waste-to-energy plants." The use of corn to produce methanol should cease, and the Jatropha (ya-tr-fa) plant (diesel bean) should replace the use of corn for an alternative energy. The Jatropha plant should be genetically modified to produce bigger diesel seeds. More geothermal plants should be built to generate electricity on land and under the sea. And in cities close to a natural gas supply, automobiles should be powered with natural gas. In addition to the focus on energy, volcanoes can be and should be conquered so that heavy populated areas will no longer become covered by lava and ash. Hurricanes can be controlled and the terrible destruction wrought by these monsters will be history. This must be a government project for the method is a mighty undertaking.

Prophecy Given: A *new computer* will be introduced that will be more efficient and user friendly beyond anyone's dreams as the 21st Century begins. Finally, in the latter part of the 21st Century, *gravity will be neutralized and engines* will be invented that will propel space ships at thousands of miles per second. A great future is coming for technological developed nations.

**LET US BEGIN OUR JOURNEY INTO THE SCIENCE OF THE TWENTY-FIRST CENTURY.**

Before we begin, we must crack the shell of accepted science of the 20th Century. Therefore, since I have made it very clear to all interested scholars that the concept of inertia being the force that enables planets to revolve around the Sun is a false idea, let us review the property of inertia that is inherent to matter. The first scientist to challenge the dictum of Aristotle concerning moving mass was the French scientist Jean Buridan (1300 –1385). Buridan correctly postulated that mass in motion had an inherent moving property that kept the mass moving. This was the first known case where a scientist introduced the concept of inertia.

Three centuries later, Isaac Newton formulated his three laws of motion. Newton probably knew of Buridan's concept and was preceded by other notable scientists that studied motion. These were: John Wallis (1616 –1703), Christian Hygens, and Sir Christopher Wren, (16322 –1723.) Whether Newton knew of Buridan or not, Newton's work forever established the concept of inertia.

Newton's first law of motion established the concept of inertia when he stated that all matter or mass possesses a property of inertia. This property is divided into two states: stationary and moving. Thus Newton's first law states: "a body at rest will remain at rest until acted upon by an outside force and a body in motion will remain in motion until acted upon by an outside force." This concept is one of the key or fundamental principles of Einstein's theory of general relativity." Einstein erroneously claimed that the planets revolve around the Sun by inertia and since they (according to Einstein) do not meet any motional resistance, they continue to revolve by reason of their motional or moving inertia.

With this given history and the explanation about Einstein incorporating the inertia idea into his theory, we are ready to proceed. Therefore, just honestly answer the first four very simple questions of the sixteen that follow and we are on our way into the science of the 21st Century.

Question 1 pertains to the force that causes the planets to revolve about the Sun, it has two parts: first part: do the planets revolve by inertia as Einstein claimed?

Before considering the second part, first consider Kepler's concept, for he stated that "there exists only one moving Soul in the center of all the orbits; that is the Sun, which drives the planet the more vigorously the closer the planet is"—this force becomes weakened— "when acting on the outer planets because of the long distance and the weakening of the force which it entails."

My modern-day paraphrase of Kepler's statement is: The rotating Sun holds the planets in a gyroscopic, rotary, gravitational force field and drives the planets around the Sun in this extended force field. Kepler's quote was revealed to me in the 1968 Putnam book by Robert Silverberg, *Four Men Who Changed the Universe*.

Before you answer this question as to whether the planets revolve around the Sun by inertia or whether the rotating Sun holds the planets in a gyroscopic, rotary, gravitational force field and drives the planets within their orbits, read the next paragraph.

The Sun rotates in a counterclockwise direction, and all the planets revolve in a counter clockwise direction with mechanical precision according to Kepler's third law. Kepler's third law states, "the distance of each planet from the Sun cubed, equals its period squared, $d^3 = p^2$."

Question 2: Do these facts given in the preceding paragraphs suggest greater creditability be given to Kepler's concept where a rotary gravity field drives the planets? Or, should greater credibility be given to Einstein's postulates that states that the planets revolve in an "inherent curvature of space" due to space's "non linearity," revolving in a "Space Time Continuum" "by inertia" "taking the shortest possible path in a warped and deformed space" where "gravity is a condition of space in the vicinity of matter"?

Question 3 is a pointed question to the preceding question. Is gravity a force radiated by all matter [mass] or is gravity a "condition of space, in the vicinity of matter," as Einstein claimed? Assuming that you have selected your answers to the three questions concerning the accepted solar system dynamics, then answer question four's two-part question that is pertinent to Mercury's orbit.

Einstein claimed: the revolution of the ellipse of Mercury is a "relativistic effect."

QUESTION 4, Part 1: *Are* the planet's orbits actually proper ellipses?

Part 2: If the points of aphelion and perihelion in Mercury's elongated orbit do revolve, is this advancement in the direction of Mercury's motion a "relativistic effect?"

If you are not able to defend the answers that you have chosen, then physics and astronomy of present day state of the art should have ready answers.

Within the entire body of accepted science in this Space Age, answers to the four given questions should also be readily available. Therefore, a defense of presently held concepts should be sufficient to explain the reasoning for the answers to the four given questions. For remember, doctorates—that is a PhD, is given to physicists and astronomers by including general relativity quotes in their thesis.

While the questions given are plain and uncomplicated, truthful answers to the given four questions are denied to that entire search within concepts of present held science. As incredible as it seems, the truth to these four questions cannot be found in any scientific text, for present-day physics and astronomy cannot or will not truly verify that Kepler is correct and Einstein is incorrect; nor the converse, that Einstein is correct and Kepler is incorrect. It is quite a weird situation for a science that is supposed to teach truth and be respected.

While this conflict of concepts that pertains to the dynamics of the solar system does exist, the concept is not addressed by any respected text. The reason for this is because physics books, astronomy books, and encyclopedias do not truly explain why the planets revolve as they do, excepting to falsely attribute the planet's revolving motion to inertia.

I have tried, when talking to fellow scientists on a one-on-one discussion, to elicit from them a rationale as to why they believe that planets revolve by inertia, and every time that I have tried to pin down a scientist concerning the motions of the planets, they escape into relativistic vernacular, eventually attributing planetary motion to inertia, even though known facts that are clearly explained in this volume refute the idea that the planets revolve around the Sun by inertia. Why today's scientists respond with this false concept is not only *weird* but is a mystery, *unless* organized science is ashamed to admit that they have been teaching false concepts for decades, yes decades, about ten decades, almost a hundred years.

It is general knowledge that Einstein was never awarded a Nobel Prize for his theory of general relativity because some of the greatest and respected scientists in the world claimed that the correctness of Einstein's theory rested on beliefs—that is, opinions, and not proven facts.

In addition, no one should ever forget that Newton's laws

derived from Kepler's laws are used to place satellites into orbit and have sent men to the Moon.

It is now March 2, 2005, it has been over 12 years since I started this volume, and I have edited and re-edited it many times. At this time, I wish to include a grossly inaccurate article that was posted on a website in 2005 by the BBC. Their site: BBC, Science & Nature, Space, the Sun, presented some very interesting data about our Sun. However, three paragraphs under the title LOCAL HISTORY contained a grossly inaccurate history about our solar system. The site has since been pulled from the Net, although I copied the three paragraphs. They follow:

> "The Sun has been given many names over the course of history. The Greeks named it "Helios," the origin adjective "heliocentric" (meaning centered around the Sun). The Romans referred to the Sun as "sol." Until the Middle Ages it was assumed that the Sun orbited the Earth. The first man that received public attention from his theoretical work suggesting that the Earth revolved around the Sun was Nicolaus Copernicus in the early 16th century. However, his view of the Solar System wasn't accepted for many years until Newton formulated his 'laws of motion.'"—*BBC NEWS*

That is a grossly inaccurate statement by the BBC, for Johannes Kepler submitted the three laws that *proved* the Sun is the center of the solar system. Copernicus probably learned of the heliocentric nature of our solar system while studying the works of ancient Greek astronomers, and this generalized concept of Copernicus was organized in a book by a young associate of his by the name of Rheticus. A detailed history of this is given in the volume *Music of the Spheres*. Kepler was not even mentioned in the BBC website.

When this work is published, the BBC will make a correction as will all physics and astronomy departments in every school of learning.

Let us focus our attention to Kepler's laws and prove to the world that the BBC is wrong In addition to proving the BBC wrong, Kepler's hypothesis concerning the solar system dynamics is again elevated to its proper stature—that is, the BBC and Einstein are both proven wrong.

Kepler's second law states that a planet moves faster closer to the Sun, and it moves slower as it moves farther from the Sun. Also, a planet moves through equal areas of space in equal times. Thus, any time period within a planet's orbit, when multiplied by the area it encompassed, the quotient is always equal.

KEPLER'S 2ND LAW

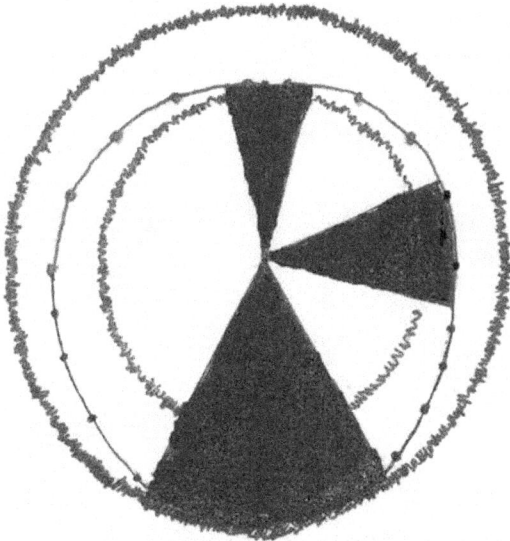

A planet's orbit is an uneven oval,
that has characteristics of an ellipse.

To see this law working with fine animation, go to Google. com and do a search for "Kepler's Laws with Animation."

The fact that a planet moves through equal areas of space in equal times enabled the concept of the planet's orbits to be accepted as an ellipse. The mathematical fact also applies to the planet's true, uneven oval orbits. Kepler's third law states, "The squares of the periods of two planets are proportional to the cubes of their mean distance." His third law proved that a binding force drives the planets around the Sun in mechanical precision. This fact, coupled with his second law, rules out chance, happenstance or coincidence as the factor governing planetary revolving motion.

**INERTIA AS THE DRIVING FORCE THAT CAUSES THE PLANETS TO REVOLVE IS REFUTED.**

By including the fact that planets revolve in the same direction as the Sun rotates and obey Kepler's third law, Einstein's idea of attributing inertia to be the driving force governing the planet's revolving motion, does not have any logical support. Kepler's third law illustrates harmony of the worlds in mathematical form. It is illustrated in the following data. While Kepler knew the rotating Sun drove the planets around the Sun when he departed this Earth, if anyone wonders why the distance cubed does not exactly equal the distance squared to the finest increment, with Mercury, Venus, Jupiter, Uranus, and Pluto, the answer is tied to those of the Observatory that made the measurement and when they made the measurement. We must depend upon those that took the measurements and as proven, data from the Jet Propulsion Lab and the Naval Observatory does not always agree, so it is not a mystery.

To see Kepler's third law given in the given mathematical data leaves no doubt that the planets are driven around the solar system by the rotating Sun. However, while Kepler knew the rotating Sun drove the planets around the Sun, when he departed in 1630, there were still remaining unsolved mysteries of the solar system.

Still remaining was lack of understanding as to why the planets revolve in a plane known as the ecliptic plane, (see Aust and Wake) and why the orbits were not true ellipses but were uneven ovals (see Aust and Wake), what if any were the geometric relationships in the revolving planets that influenced human behavior.

These puzzles remained for 328 years after Kepler departed until this reclusive researcher yearned to be a vehicle and servant of God as the riddles of the universe were sought. Now, in 1997, it is 367 years after Kepler and a great new period is about to unfold. Kepler was first trained to become a minister, so his moorings to God our heavenly father were deep and fast. Certainly he struggled, for man will advance by the sweat of his brow, but Kepler qualified to serve God, and the laws of the solar system came through Kepler, not those that defied God.

Kepler's second and third laws established modern astronomy and forever ended the idea of an Earth-centered solar system.

Isaac Newton's law of universal gravitation with the Cavendish constant advanced Kepler's second law to great precision. Modern scientists accept the fact that every planet began to revolve in a counterclockwise direction, in the same direction as the Sun rotates, and no planet revolves in a clockwise direction. However, there is no agreement as to why.

KEPLER'S THIRD LAW
$$d^3 = p^2$$
$$d^3/p^2 = 1$$
PLANETARY DATA BELOW

| | orbital semi-major axis in Astronomical Units | orbital time in Earth days & Earth years | |
|---|---|---|---|
| Mercury | .387 | 87.97 .241 | 1.002 |
| Venus | .723 | 224.7 .615 | 1.001 |
| Earth | 1.000 | 365.25 1.000 | 1.000 |
| Mars | 1.524 | 686.99 1.881 | 1.000 |
| Jupiter | 5.203 | 4,331.865 1.86 | .999 |
| Saturn | 9.539 | 10,759.169 29.46 | 1.000 |
| Uranus | 19.19 | 30,685.38 84.01 | .999 |
| Neptune | 30.06 | 60,191.37 164.8 | 1.000 |
| Pluto | 39.53 | 90,472.42 248.6 | 1.001 |

In order to understand *why* the planets continue to revolve in an orbital path and *why* each planet's path is an uneven oval requires a deeper scrutiny. Since Mercury, Venus, and the Earth, definitely have some type of a force holding them in orbit; otherwise, they would have been pulled into the Sun shortly after they were created. Since the first three terrestrials are small rocks and the Sun contains over 98 percent of the mass of the solar system, something maintains their stable position. But the three inner planets, indeed all the planets, continue to revolve around the Sun as though their own gravitational force would hold them in place.

This is a preposterous idea since Mercury is a small rock and should have been pulled into the Sun at some time as the Sun gases contracted *unless there is another reason to explain the planets stable positions.*

Kepler may not have been the first to ponder this question, but his deep perception enabled him to correctly attribute the power of the rotating Sun to be the force that *held* the planets in their orbits and was the *driving force* that caused the planets to revolve.

Einstein wrestled with the question as to why the inner planets were not pulled into the Sun and why all the planets revolve as they do. Within Einstein's analysis, he became satisfied with a Buridan-Newton concept of inertia and from this he created an idea of a warped and deformed space,'" where the planets revolved in a trough. This theoretical conjecture or imagination seemed reasonable to Einstein IF the planets revolved by inertia.

The idea of a fixed orbital confine is not ruled out and you will discover more about this as you read. But attributing inertia as the driving force does not take into account three critical factors. First: the planet's revolving direction, (revolving the same direction as the Sun rotates). Second: their precision rate of movements. Third: the planet's latitude.

These three critical factors were never taken into consideration in Einstein's theory of general relativity, and that great deficiency in general relativity is part of the root reason why Einstein never was awarded a Nobel Prize. The Nobel Prize review board stated that Einstein's "Theory of General Relativity" was based upon opinions and is not supported by facts.

In addition to being beyond probability for all the planets to have begun to revolve with precision movement according to their distance from the Sun, another factor about the latitude position of the revolving planets must be considered, for *all* the planets revolve in the Sun's rotary gyroscopic, gravitational force field that extends from the Sun's greatest girth—that is the Sun's zero line of latitude.

In consideration to the three critical factors involved with planetary motion, the idea of attributing inertia to planetary motion, completely fails.

In summary, these three facts about the planet's movements are again given.

Their direction of movement - // all planets revolve in a rotary gravitational field that is counter clock wise, that is, the planets revolve in the same direction as the Sun rotates, // the orbital movement of each planet is so precise that each planet's distance from the Sun determines its orbital rate of movement so that their $d^3 = p^2$ // their latitude position within the Sun's gyroscopic gravitational force field—that is, each planet revolves in nearly a flat planed area that extends out from the Sun's zero line of latitude. The plane of the planet's revolution that extends outward from the Sun's greatest girth is narrow; its near flat plane extends to about a total of seven degrees from the Sun's central girth. However, the Aust expands vertically and horizontally in the region past Neptune. This causes Pluto to have the greatest eccentricity and the greatest tilt to the ecliptic, a little over 17 degrees.

These three given conditions *could not* have occurred by happenstance; therefore, in respect to honesty, when these three facts are considered together, it proves conclusively that Einstein's concept of inertia being the driving force for the planet's revolving motion must be completely dismissed.

Please keep in mind Einstein's claim. He claimed that

planets revolve in an "inherent curvature of space" due to space's "non linearity," revolving in a "Space Time Continuum" "by inertia," "taking the shortest possible path in a warped and deformed space" where "gravity is a condition of space in the vicinity of matter."

The idea that planets revolve by inertia in non-linear space taking the shortest possible path in a space-time continuum may be great for science fiction, but the idea *Is Not* supported by facts. For these powerful reasons, general relativity is revealed to be a non sequitur theory—(AHD "conclusion that does not follow from the premises or evidence.")

That concludes the basic facts concerning solar system motion.

I have tried to present the vital facts in an uncomplicated manner. The motions within the solar system will be considered again. For now, I'm sure that you the reader were able to peer into the abstract principles; you are ready to proceed with a deeper study.

Today is Thursday May 5, 2011. On the front page of *The Wall Street Journal*, on the left side where articles pertaining to Business & Finance // Worldwide, under the general heading of What's News, at the very bottom of the column Worldwide, a notation caught my attention. "*Scientists ended* a 52 year, $750 million space-physics experiment that affirmed Einstein's theory of relativity. A3" At the top of the inset showing graphics to depict how measurements were used to prove Einstein's theory of general relativity, a further statement claimed "space and time are woven together in a four-dimensional fabric that can be warped by massive objects, such as Earth," reported the Wall Street Journal.

Three scientists working at Stanford University (that is one of the universities that was not interested in my work that proves astrology) are so enamored by the relativistic jargon that they actually believe in a "space-time continuum." The term *continuum* is a mathematical term that is supposed to insulate the concept of duration with the James Clerk Maxwell's formula of electromagnetic radiation. It fooled thousands of scientists since Einstein introduced his laced together/woven together concepts of previous scientists into a gobbledygook theory.

It is true that mass in space can affect and be affected by other mass bodies. However the rotating Sun creates planetary orbits and controls all the space in its gravitational domain. Interplanetary gravitation can slightly alter the motion of nearby mass-bodies, according to Newton's law of universal gravitation.

The 52 years of work and the $750 million dollars that was spent proved Isaac Newton's law of universal attraction and did *Not* prove any ideas built into Einstein's general theory.

The scientists and technicians that worked on this 52-year project did not know:

1. The so-called law of conservation of energy is only one half correct, for energy is not always conserved for energy can be dissolved or rendered inaccessible or its capacity to perform work is destroyed.

2. Planets do not revolve around the Sun by inertia; planets are driven around the Sun by the Sun's gyroscopic, rotary gravity field.

3. Magnetism is not a condition of space in the vicinity of matter; magnetism is a flowing force field with two poles. A magnetic force flows into the Earth's magnetic South Pole and exits from the Earth's magnetic North Pole.

These three facts were not known by the three researchers at Stanford's research facility; therefore, their effort to prove general relativity was doomed from the beginning.

Einstein's theory of general relativity is a sterile theory; the ideas woven into his theory cannot produce anything practical. General relativity is revealed to be a non sequitur theory whose conclusion does not follow from the premises or evidence.

My pragmatic nature yearns to have volcanoes and hurricanes controlled and science books changed to explain to future generations how the theory of general relativity duped the entire scientific world for about 100 years.

**CONTINUING WITH THE MAIN BODY OF THE SCIENTIFIC EXPOSÉ THAT I WROTE BEFORE 1995 AND AFTER 1995.**

Students of science, and all my readers that are interested in truth, this volume was posted on the web for several years. In that time, hundreds of scientists that were working in academia, planetariums, and scientific organizations, plus science students have read the facts that are presented in this volume and not one person has answered the questions that are presented in this work. I have contacted some by phone and others by email and have asked them to please read my web pages and please answer the questions that are given. I have received promises from my contacts that my questions will be answered and that they will get back to me. These promises were untruths, or may I use the word *lies*, for not one person has ever answered the questions.

I truly believe the reason for their refusal to answer the questions, is a personal fear. If the questions were answered honestly, then Einstein's theory of general relativity would be affirmed to be false. Thus, their position in organized science would have been tainted, for basic foundation concepts in the entire body of physics are founded on false, unproven postulates or ideas. This is an almost unbelievable and fantastic claim, so this work is presented in a manner to supply evidence to these claims and not to simply present a postulate based upon unproven ideas. Thus, the reasons for each premise of Einstein's theory of general relativity are examined in depth. Also, I present counter-propositions based on facts to each false premise of Einstein's and reveal how Einstein came to believe the propositions presented in his theory of general relativity.

Also, in addition to the dynamics of the solar system not being carefully explained by present science, there are questions concerning particles and energy that science does not conclusively explain, even though there is ponderous evidence that explains the nature of electrons and energy.

While this claim may seem to you as being unsupported, I draw to your attention the fact that electrons are known to be negative charged particles with infinitesimal mass. According to the latest scientific research, the mass of an electron is 1/1836th of a proton. On the basis of this knowledge, I ask you to answer the three following questions concerning electrons.

Question 1: What is the inner nature of an electron? Einstein already ventured into this realm of theoretical physics, for he speculated that "perhaps the electron has an inner action"_ and I ask you, does the electron have an inner action? That question produces two more questions to gain greater understanding for if an electron does not have an inner action then question two must be answered.

Question 2: What causes electrons to disappear when fired into a thin metallic foil and concentric circles of interference patterns emerge? There remains the third question concerning an electrons charge.

Question 3: What gives an electron a negative charge?

If you have answers to these three questions, hold your answers, for electrons are energy particles and you should be able to define and explain the nature of energy. Do not allow this deep probing to bog down your reasoning, for great minds have wrestled with understanding the nature of energy, and from these theoretical conjectures a universal concept has been accepted but not proven. Before you suspect that I may be misleading you, please consider the theoretical concept or postulate of two German physicians, who became interested in physics J. R. Mayer (1814 –1878) and Hermann Helmholtz (1821 –1894). After much meditation on the subject of energy, Mayer came to believe that the basic energy of the universe is fixed and therefore cannot be created by man or destroyed. He presented this idea in 1842 from his meditations on the subject, but the most important requirement in postulates was missing, for he did not present a verifiable test to prove his suspicions.

Then later the second prominent German physician/physicist, Hermann Ludwig Ferdinand von Helmholtz, speculated on the nature of energy and arrived at the same erroneous conclusion as Mayer—that is, that energy cannot be destroyed. In 1847, Helmholtz submitted his idea, and like Mayer did not present a test to prove his concept.

The idea that energy cannot be destroyed was accepted by organized science on the basis that Helmholtz is a smart man; therefore, it must be true. However, later other prominent physicists accepted this idea. Max Planck's erroneous idea that light energy can be stored in a black box misled other physicists and the law of conservation of energy more firmly became fixed in physics as a law just as Ptolemy's Earth-centered solar system was accepted by science and the Catholic Church for hundreds of years.

When Albert Einstein was born into the world, energy's conservation concept was accepted as the gospel truth, and Einstein never analyzed the claim sufficiently to challenge the claim. Therefore, Mayer, Helmholtz, Planck, and others, plus Einstein, incorporated the fallacious concept known as energy's conservation principle into theoretical conjectures. From this claim by many prestigious physicists that followed Mayer and Helmholtz, I ask you the reader two questions.

Question 1: Has the conservation concept been proven?

Question 2: Or has this concept been accepted on faith?

Before you give a conclusive answer to these questions concerning energy, the following question that is submitted in capital letters will assist you in being able to answer with clarity and truth.

QUESTION 3: WHAT HAPPENS TO BLUE LIGHT THAT IS DIRECTED FOR AN EXTENDED PERIOD OF TIME THROUGH A FIBER OPTIC TUBE OR TUBES OR WITHOUT FIBER OPTIC TUBES THROUGH AN OPENING, INTO A CLOSED BOX THAT HAS BEEN SPECIALLY DESIGNED TO PREVENT ANY LIGHT FROM ESCAPING, AND IS DESIGNED TO BREAK UP THE LIGHT WAVE?

I ask you to clearly understand that the special box is lined with near perfect fiber optic mirrors with diffusion glass on top of the bottom mirror. Before you formulate your answer, keep in mind the box being described is a closed, mirrored box, except for an opening to allow the blue light to enter. Assuming that you have selected an answer according to accepted scientific principles, let's review the properties of light, and then allow your answer to be cross-examined by the questions and conclusions reached in the next paragraphs.

**UNDERSTANDING LIGHT**

Men and women of science: It is now 2017 and I am 89 years old. I am making this special address to you people since it is your duty to teach your students truth about the environment in which we live. I am sorry to say, in the last over fifty years I only ever met one physicist that was honest. I have been laughed at, lied to, and regarded as a person that lacks comprehension. My claiming that Einstein's theory of general relativity is a laced-together idea with diverse concepts that is based upon false assumptions was greeted—to use a metaphor—as a door slammed in my face. I was regarded as a person not to be taken seriously.

In addition, when I stated that astrology is a true phenomenon, the respect that I received from organized science was zero. Now in 2017, over two generations of physicists have come and gone since 1960 and a new mentality or a mental scope is leading the body of organized science. I entreat all members of this third generation of scientist to be brave in this new era. Decades have passed since I began an effort to convince science about the errors and falsehoods in the body of physics. So read the following expose of light and please do not change the subject as physicists have done in the past and do not begin to offer counter ideas based upon opinions. The entire exposé of light has a goal, and that is to use the known facts about light that concludes with the scattering of light and the interference patterns created when light is scattered. So—let's get going.

## THE NATURE OF LIGHT

There are many electromagnetic radiations in nature; they are given this general name because there are two forces that are inherent to all these radiations. These forces are electric and magnetic. In addition, each electromagnetic radiation has complex energy concentrations that are continually changing in energy content as the radiation advances.

Light is an electromagnetic radiation; it travels in a constant velocity slightly more than 186,282 miles per second in free, open space just as all other electromagnetic radiations. The first recorded attempt to measure the velocity of light was made in 1676 by the Dutch Astronomer Roemer Olaus (1644–1710). His measurement was incorrect (his figure was 227,000 kp–141,056.217 mps), but he did reveal that light traveled at an astonishing velocity.

The next in the line of development is an Englishman by the name of Thomas Young (1773–1829). Young was a very versatile genius, becoming a physicist and a physician. It was Thomas Young's work that produced the proof of the wave nature of light. He proved the wave motion of light was a transverse wave. That is, oscillations of the wave were a back-and-forth movement, or to paraphrase, the oscillations were 90 degrees to the forward direction of the light. Next in line as we follow the development of the understanding of light is James Clerk Maxwell (1831–1879), who gave us the formula of expanding shells (later improved upon by Lorentz).

The velocity of light was still a topic discussed by researching scientists, and a German-American by the name of Albert Abraham Michelson (1852–1931,) arrived at a near accurate figure in 1882. His parents immigrated to the states when the young German boy was two years old.

In 1923, Michelson again tried to accurately measure the velocity with more sensitive equipment and arrived at the figure of 186,271 mps. By the 1940s the accepted figure was 186,281 mps, now accepted to be 186,282 mps. However, this constant velocity only pertains to light traveling in free, open space, for light travels at varying rates of movement as it passes through different transmission mediums.

In experiments at Stanford University, Kaiserslautern University in Germany, Harvard-Smithsonian Center for Astrophysics in Cambridge, and at the Hanscom Air Force Base near Boston, light has been slowed to at first a walk speed then slowed more until light was completely stopped (*Science News*—January 27, 2001).

Light travels in waves, but can be polarized and thereby the polarization alters its wave geometry without changing its velocity. Light expands as it advances, but can be focused into a coherent beam as a laser light. Also, light advances by going through seven major and distinct steps and innumerable connective steps between. As light advances, it changes its frequency from red through the colors to violet, then through the colors again in reverse order and returns again to red. This repeating change is referred to as a "cycle." When viewed, this continually expanding seven-step cycle is known as the "light spectrum." We see these seven distinct steps as colors, most notably in a rainbow.

Light can be divided into its seven spectral colors, as Isaac Newton demonstrated with a glass prism. The total light spectrum contains regions in the violet end and the red end that are invisible. Light's repeating seven-step cycle is perpetual and could exist for eternity if the light wave never met with a disturbance to its interactions. Light can be refracted—that is, it slightly changes its direction when it passes through mediums of different density. This causes a slight visual displacement of objects when viewed through more than one medium at the same time.

In 1926, the minute concentrations of energy in the light wave structure, was given the name of Photons by the American chemist Gilbert Lewis (October 23, 1875–March 23, 1946). Organized science accepts the proposition that photon particles are indestructible. I proved that this idea is not true and submit the fact that photons are not stable outside of the light wave. Photons should not be confused with subatomic particles, for subatomic particles are stable in free space. Photons are referred to as particles sometimes rather than using the long-winded term *"concentrations of energy."* It is the photon's positions in the electromagnetic wave's interactions that give these minute energy concentrations stability. Before Dr. Gilbert Lewis, Dr. Compton (September 10, 1892–March 15, 1962) and Max Plank (April 23, 1858–October 4, 1947) the term photon, was originally suggested by Isaac Newton in 1704, as minute concentrations of energy that he named *corpuscles.* These minute concentrations of energy in the light wave are being continually altered in energy concentration, or quanta of energy as the wave advances.

However, at all times the energy in the total wave remains constant (excepting when interference patterns have cancelled some vibrations) and light-wave energy at any point on the wave is inversely proportional to the frequency of the wave.

Light can be reflected uniformly by smooth reflective surfaces and can pass through some matter as pure glass and water. In the case of water, the ability of light to pass through water is also dependent upon the water's purity. Even if the water is pure, light will lose its intensity as the distance increases.

Light can be absorbed by some surfaces and absorbed internally by certain types of matter as an electro-luminescent material or photoelectric cells. The absorption of light into matter can cause chemical, electrical, or physical changes to occur to the matter or physical substance, dependent upon the nature of the substance.

In the violet end of the light spectrum, light vibrations merge into ultraviolet light. In this region of the light wave, the light has characteristics that are vastly different to light at the opposite end or red end of the spectrum. Violet light and blue light contain no photon particles that produce increased thermal motion or heat. The chief property of violet light is its ability to cause chemical changes as bleaching and slightly into the ultraviolet range to kill bacteria. Also, the higher vibrations of the violet light where violet light merges with ultraviolet does damage human skin thereby causing melanoma or skin cancer.

The red end of the spectrum merges with the infrared region and in this region the Photon particles are known as infrared particles. When infrared particles contact matter, heat is generated (thermal motion) in the substance that absorbs the infrared particles.

Infrared particles can cause physical changes to occur in the absorbing substance, as raising the temperature to the kindling point and causing rapid oxidation or causing the substance to burst into flames. A clear magnifying glass concentrating the Sun's rays onto paper until it bursts into flames is a good illustration of the power of infrared particles in light.

Light can also be filtered. If sunlight or artificial light is directed to pass through colored glass, only the vibrations akin to each color in the light wave will pass through the colored glass; all other vibrations or colors of the light wave will not be able to pass through the colored glass.

Thus, if blue glass is placed over a magnifying glass when the sunlight is concentrated on paper, not only will the infrared vibrations and red color be filtered from the light wave but all colors below blue and violet will be filtered from the light wave. Consequently, the paper will not ignite into flames. The blue colored glass will become warm, but the paper on which the blue light is focused will not become warm. If pure sunlight is concentrated by a magnifying glass onto your hand, in a few seconds the heat will become unbearable.

If a blue or violet light filter is placed over the magnifying glass, a blue dot of concentrated light will appear on your hand. However, it will not get hot or will not even get warm, because no infrared particles are present in blue light.

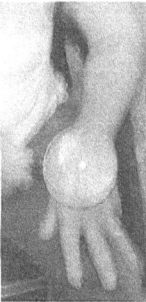

To the left:   Magnified sun light will burn skin, will cause paper to ignite into flames.

To the right: magnified sun light with a blue or violet light filter will not burn the skin and cannot ignite paper.

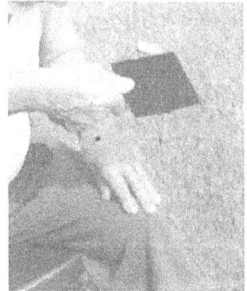

PLEASE READ THE FOLLOWING CAREFULLY, FOR PHYSICS TEACHES A FALSE IDEA TO BE A FACT. THE FOLLOWING TEST PROVES PHYSICS TO BE WRONG.

..............................................................

# Light is not always uniformly reflected by the surface of matter for light can be diffused by rough surfaces. *Diffused light can be scattered* and when light is scattered, the light wave structure departs from a uniform movement and is caused to change direction randomly due to collisions of the light particles i.e. Photons.

..............................................................

Diffused light can be scattered and thereby this causes an interference in the light wave structure. This scattering of light in a confined enclosure causes a 'breaking up' of the light wave structure due to interference. This is the critical fact in the nature of light that proves energy dissolves. Please read this paragraph slowly and allow the truth to burn into your memory. The condition of interference causes some vibrations or frequencies to be reinforced and others to be cancelled or destroyed.

Therefore a reader, by capitalizing on this characteristic of light, light is not always uniformly reflected by the surface of matter for light can be diffused by rough surfaces. The condition of a controlled environment can be created to purposely cause interference to continually cancel, destroy or dissolve the light wave.

## TO ALL PHYSICISTS AND EVERYONE WITH A REASONABLE MIND.

Please read slowly, for the interference characteristic of light has enabled me to create controlled conditions that prove the dissolving nature of light.

Since your understanding of the light wave has now been refreshed, you must consider an attribute that organized science claims is inherent to the light wave. Light is supposed to possess an energy that is indestructible since science teaches that energy cannot be destroyed.

Within the honest search for truth, the following four questions must be answered.

Question 1: When blue light that does not contain any infrared particles is directed into a special constructed box that is lined with mirrors and some diffusion glass, does the light exist in an eternal state within the box?

Question 2: Or does the light break all the laws of physics and transform into heat?

Question 3: Or does the light transform into elections within the box?

Question 4: Or does the light dissolve into a primal energy state? If the blue light exists within an eternal state, it will jump out at you when you open the box. How absurd to even consider this possibility. If it transforms into heat, a thermometer can reveal this; if it transforms into electrons a Geiger counter can reveal this.

Remember: light is energy, and energy, according to age-old beliefs—cannot be destroyed.

# WORLD SCIENTISTS: HARK TO THE NEXT PARAGRAPH.

This test should be considered by all scientists with the utmost interest because if the energy is lost, a foundation theoretical proposition of all physical sciences which is called the law of conservation of energy is proven to be invalid or to be a false concept. This creates an entirely new concept, and physics will have to be reshaped to agree with truth or agree with reality.

Also, the first law of thermodynamics is the law of conservation of energy, and it is being proven to be a fallacious concept. In addition to refuting the conservation of energy idea, a basic concept of Einstein's theory of general relativity is proven to be false. Therefore, this negates two propositions in Einstein's theory of general relativity. The first considered is the proposition that planets travel by inertia. The second proposition is the idea that energy cannot be destroyed. In this analysis, there are more propositions in Einstein's theory to be proven false. When the analysis is finished, only remaining will be the claim that physical objects cannot exceed the speed of light.

Among all the revelations submitted in *My Search for Truth*, the greatest explosion to physical science will be proving astrology. Although, the proof of astrology is in Volume II and this is Volume I.

Therefore, I believe the second great explosion will be proving that energy dissolves. However, rather than make a personal claim, allow me to present the present held ideas from an encyclopedia. Infopedia 2.0 states,) "The law, which states that the sum of kinetic energy, potential energy, and thermal energy in a closed system remains constant, is now generally known as the first law of thermodynamics )E (q.v.). In classical mechanics, the fundamental laws are the laws of conservation of linear momentum and of angular momentum (see Mechanics; Momentum ). Also fundamental is the conservation law for electric charge."

Come, come, come gentlemen, be honest. The truth is resolute: every change in mass and energy must occur within an environment and the environment always extracts or takes some energy from the change. Otherwise you gentlemen could create a perpetual motion machine.

## HONEST SCIENTISTS WILL BUILD A MIRRORED BOX AS DESCRIBED.

Scientists working in labs can build a much larger box than I built. Even a 4 x 4 x 4 box (or larger) can be built solely for the purpose of having enough space for a carefully constructed opening for great quantas of blue light to be directed into the box. Instruments more sensitive than the ones that I used to measure the amount of energy that goes into the box and a very sensitive heat detector are also available in scientific labs. In addition, an initial receiving mirror on the inside of the box is needed to reflect the light downward to diffusion glass.

Temperature probe on the inside. dial on the outside (carefully caulked)

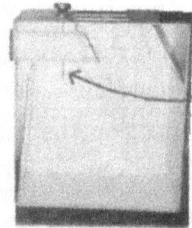

The box that I built is on a small scale, although it was large enough to prove that energy dissolves. I used my box on several occasions in my office and study. The results of my tests, by using mirrors with some diffusion glass in a box with blue light, prove that energy dissolves.

Actual photographs of the device that I built explain the necessary conditions for every concerned scientist who is interested in building an energy-dissolving box for their home or the physics classroom. Also in this volume, there are five additional concepts that are submitted that will enable a new physics and a new theoretical physics to be established. Three of these concepts can be proven by tests, and these tests are also submitted.

For all eager experimenters that wish to demonstrate the energy-dissolving attribute, I suggest that you build two large boxes, one lined with photoelectric cells, the other lined with mirrors and some diffusion glass. I placed the diffusion glass on top of the bottom mirror to diffuse or break up the light wave.

The second box that is lined with photoelectric cells can use the electrical energy to drive a toy train or another mechanical device. The box lined with mirrors and diffusion glass will not change the light energy into another form of energy; the light energy will be dissolved.

In the box that I built to prove that energy dissolves, I adhered to scientific principles and rigorously controlled conditions.

Temperature probe on the inside.
dial on the outside
(carefully caulked)

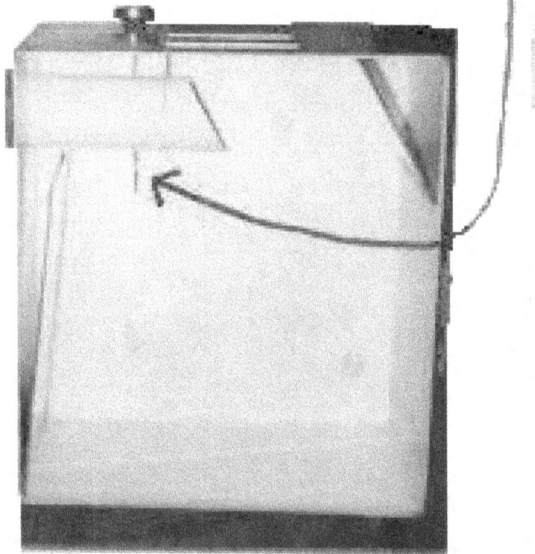

I have included three pictures of the Energy Dissolving Box that I built. You could use your bathroom or an ordinary shoe box or a black box to show that energy dissolves, but science demands exact measurements, and with my box I have provided exact measurements. Every question that comes to your mind concerning the blue light's place in the spectral distribution and the blue light filter will be answered, so please be patient and follow closely the examination of scientific principles and the development to the detailed account of the Dissolving Energy Box.

The pictures given illustrate the design simplicity. The diffusion glass that covers the bottom of the box can be seen in the right of center picture.

There are two pertinent questions concerning the possibility of any transformation of blue light into another energy form and one question regarding the disappearance of the blue light. These questions I again repeat.

Question 1: Did the blue light transform into heat?

Question 2: Did the blue light transform into electrons?

Question 3: Did the blue light dissolve into a primal energy?

The true answer to these questions is extremely vital. Please do not allow your mind to play tricks with you. *If The Energy Has Disappeared, It Has Dissolved.*

Please confront the reality of the situation. If the energy dissolves into a primal energy state, its capacity to perform work is lost. Now the vital question must be answered. Would this loss of ability to perform work be the destruction of energy? The answer is simply energy as a quality or characteristic has truly been destroyed. If the term *destroyed* does not appeal to you, I remind you that it is the term in the so-called law of conservation of energy. Instead of using the term *destroyed* to account for the loss of light energy, I favor the term *dissolved*, which means the quality or characteristic of the light energy to "perform work" is forever lost.

Actually, the captured light within the box simply dissolved into the *ether*. ETHER is defined as a primordial field of neutrally charged particles.

Light is energy, so after capturing light in a mirrored box with some diffusion glass, to every closed mind adherent to the law of conservation of energy and the first law of thermodynamics, I ask each of you to make the light that was captured in the box perform work. To every arrogant scientist that ridiculed my work and jeered at my claim that energy dissolves and astrological forces are real, I state to each of you again; if energy cannot be destroyed, then harness the energy that went into the box and make it perform work. Eat your arrogance, gentlemen. To your shame and embarrassment, the energy in the form of blue light that went into the box is *gone*.

It is not complicated. Plain and simple, electromagnetic energy *does* have a dissolving characteristic.

To meet strict scientific requirements, the interior of an energy dissolving box was created with a controlled environment for two necessary conditions:

One - to create an enclosed mirrored environment to capture light where dark blue or violet light could be captured and could not be transformed into heat energy.

Two - to place on top of one or more mirrors, diffusion glass to break up the structure of the light wave and dissolve it.

In addition to using a dark blue or violet light filter and creating an ideal condition in a box to demonstrate the dissolving characteristic of light, my light source that was selected is known to be the coolest light so far developed, excepting black light of ultra violet light. The light meeting this requirement was General Electric's Daylight Fluorescent light tube.

I have enclosed a copy of General Electric's spectrograph of the "Spectral Energy Distribution" in manometers for the

Daylight Fluorescent light tube.

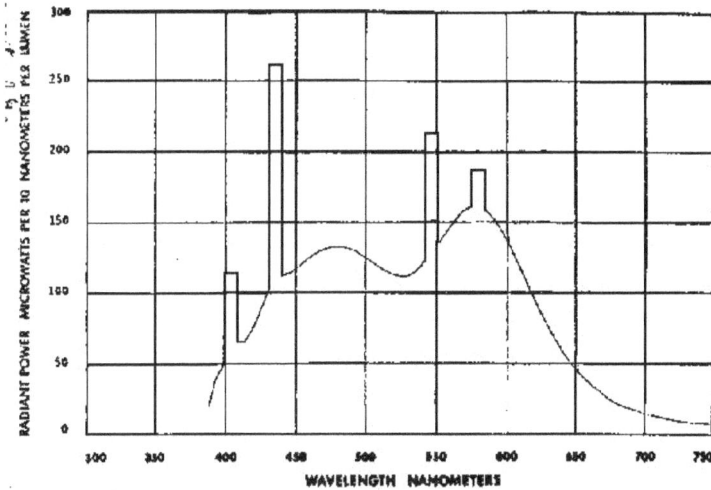

APPROXIMATE INITIAL SPECTRAL ENERGY DISTRIBUTION

I have not kept a record of each time that I had the Daylight Fluorescent tubes in operation. However, during the one period, I maintained them in constant illumination for twelve hours. During the many times I experimented, I had also had the light going into the front of the box where a large dark blue light filter was placed. During one time, a twelve-hour period passed during which I slept in my study quite near the illuminating lights.

All the energy radiated by the Daylight Fluorescent tubes was not captured in a box at that time, but due to the fact that the box was used on numerous occasions, I can safely state: the same equivalent amount of energy dissolved in the box is equal to at least a twelve-hour period. Since my areas of expertise do not include a technician's handiwork, someone better versed in these matters could supply better numerical data. I can supply some information, though, and it is sufficient to prove my claims.

Twelve 20-watt tubes were converting electrical energy into light energy for twelve hours.

12 hours x 20 watts equals 240 watts per hour -- 240 watts divided by 120 volts, equals 2 amps per second—one amp equals one coulomb—one coulomb equals 6.25 billion, billion electrons each second - 6.25 billion, billion electrons x 2 x 60 seconds per minute per hour x 12 hours x the mass of one electron (9.1 x 10-28 gram) x $c^2$ equals E If E=mc² is valid.

According to Mayer-Helmholtz, energy is indestructible, (its identity -its quality- cannot cease). And according to Einstein, the quanta of energy released into my box should have completely melted my box, or fried me in beta rays or burned me with X rays,  but the fact is, I slept quite soundly, oblivious to the energy being dissolved very near to me. My box is a crude but effective way to prove that light energy dissolves. I entreat all scientific laboratories to build a larger box that is capable of receiving more light and use more sensitive heat sensors. When the experiment is finished, your conclusion will be the same as I discovered with my small and crude box—light, and that is energy, dissolves.

Also, in some future time when the concept of dissolving energy is accepted, wattage drop or "watt loss" will be understood to be the quanta of energy dissipated and dissolved from an electrical circuit and electrons that flow through an electric current will no longer be considered as particles of energy taken from atoms that have loose electrons in their outer electron shell.

One more point is important before we continue: When mass is converted to energy a tremendous amount of energy is released as fission and fusion prove. ALSO—when gasoline or hydrogen is used in an internal combustion engine a tremendous amount of energy is released, however, the quanta of energy *does not* reach the astronomical figure postulated by Einstein's famous equation.

If that would be true, the world's yearly energy needs could be supplied by a gallon of water. Scholars, the truth follows: when you put 50 pounds of gasoline in your car and drive it until all the gas is used, the Earth becomes about 50 pounds lighter and that energy that was in the gasoline is gone forever.

Since you have read this far, scholars of science, you have entered into a depth analysis of physics and astronomy. Furthermore, if the given questions also stimulated your scientific curiosity, this work will truly be a rewarding experience for you from two perspectives, first by delving into the nature of our environment and secondly by the scientific history that is included.

These two intellectual excursions will not only be informative, but the voyage into the depth of science will produce a joy and satisfaction to all scholars of science. Let us continue but before we do I want to give you a good down-to-earth illustration that everyone can relate to that reveals with a simple example that—energy is not conserved, is spent or dissolved or is *destroyed*.

An example of energy dissolving can be illustrated by a toy train that is powered by a spring that has to be wound, and once wound, the spring has the capacity to do work. The energy in the wound spring (the potential energy) is released as kinetic energy, and the work that it performs is propelling the toy train. When the spring is unwound, the energy is not conserved, the energy is gone. To paraphrase that statement: the energy ceases to exist, it has been spent. Thus the quality or characteristic of energy, the ability to perform work ceases to exist. It isn't complicated is it? However . . .

**UNPROVEN IDEAS BEGAN TO SHAPE MODERN SCIENCE.**
Ever since 1842 and 1847 when J. R. Mayer and Hermann Helmholtz respectively, submitted their theoretical proposition concerning what they believed to be an inherent indestructibility of energy, their concept has been accepted as the gospel truth. After over 75 years of the introduction of this fallacious concept, this false idea became an integral part of Einstein's theory of general relativity. Since 1915, when Albert Einstein submitted his idea of solar dynamics, the leaders of organized science began to consider the possibility that the planets do revolve around the Sun by inertia. However, in 1919, hype from *The London Times* influenced the thinking of world physicists and general relativity was revered as the gospel truth. After the initial "jumping on the bandwagon" of general relativity, no one dared to challenge the postulates given in the conservation of energy idea, and the first postulate (accepted as a law) of thermodynamics.

To my American scientific colleagues: are you going to wait until some foreign university conducts this test then submits the finding that energy does dissolve to the world scientific community, and gets the credit till the end time? Or will an American lab be the first to receive the honors? If you procrastinate until a future researcher makes the test, how will you answer to historians when you are queried as to why you did not make the test? Will you say that American science relies heavily on opinions and not facts from tests?

I invite all to become an empiric researcher and depart from being a theoretical dreamer. Fantasizing in the science fiction realm instead of demanding evidence led to every one of you scientists to be intellectually burned by teaching general relativity to be true for over a hundred years.

**THROUGH SCIENTIFIC SEARCHES, OUR UNDERSTANDING ABOUT OUR ENVIRONMENT IS ENABLED.**

Science is defined by the AHD (American Heritage Dictionary) as: "The observation, identification, description, experimental investigation, and theoretical explanation of phenomena." However, how does the application of scientific principles affect life?

Through science we are able to understand ourselves and our environment and to harness or capitalize upon the forces of our environment. It seems quite simple; however, in order to accomplish this feat not only do we need a language, we need a writing system, a numerical system and measurement systems. These are the basic tools to apply science to our environment.

One of the basic sciences to emerge in our development was astronomy. It was a necessity to know the season-changing times or freeze to death in the winter. Then, correlating the phases of the Moon with age enabled man to measure his age with the cycle of the Moon.

We know for sure that man has been studying the starry heavens for over 10,000 years, perhaps several hundred thousand years. Every tribe in mankind's many tribes has seriously studied the starry heavens in varying degrees. As soon as records were kept, there are records of astronomers in China, tribes in the Fertile Crescent, Egyptians, and Mayans. There are probably many more whose records are lost.

As we developed, clever devices were built to make accurate measurements of the celestial bodies in our solar system and the cosmos. Astronomy has been the most studied science in the history of man, although, from about mid 1800s to the twenty-first century, physics and chemistry has replaced astronomy as the leading science. Yet within the last 10,000 years, more time and effort has been devoted to astronomy than any other science. Because many researchers in many countries have been busy with scientific research, many theories and understandings about our universe, our place in this great expanse and how we are affected by its forces, have been developed. As I edit this volume in 2003 at the age of 75, I review my life and realize that I will be counted among scientists in the past. I can truthfully say that my life as an amateur scientist never earned any money, although I loved my work as a research analyst.

In some ways, my life can be compared with Cavendish, for I worked alone in my study with only books and an ephemeris, plus the light filters. Astrology is proven not by researchers in a multi-million dollar planetarium or university lab, but by a reclusive amateur scientist that had an ephemeris as the major source of data. At first the ephemeris I used was printed. Later, the computer enabled the data to be stored on MICA floppy disks from the Navy planetarium. Eventually I bought the outstanding ephemeris named Solar Fire.

The orb structure had to come through my mind from higher consciousness. After I perceived the Aust, a new chart had to be designed that went through many phases, plus a protractor device to quickly locate Inflection Orbs. As I reflect on the development of my scientific quests, there is no doubt it was a momentous task. Why do I use self-hype in a scientific work so momentous? Because this work may never be published while I am alive, and I am having a little as I approach my 80th year of age.

I learned as a very young child that honesty is an absolute necessity in the search for truth, and I discovered very early that few people have an open mind. My lifelong Search for Truth has been wide and deep, and it never developed friendships because people generally believe in their teachings and contradicting their beliefs was a condition they could not handle.

Not only did my curiosities lead me to study the nature of our environment, but also how and why we interact with our environment as we do. Since my subconscious is constructed in a manner from my seventh chakra being more open than my other six chakras, I became more focused on how we can use our knowledge for beneficial effects. My seventh chakra being more open is a condition that was set by my birth astrological imprint. I was not really preordained for this work. I was given certain attributes or proclivities that could be developed or not developed. I had to surmount any challenges and overcome many obstacles before the openings would open in the seventh chakra.

My research about reoccurring events or cycles in the affairs of humanity led me to first discover the Naronic Cycle. There before me in 1958 I saw new laws of the solar system and in addition, as the years passed, my cross referencing data led me to other new discoveries. As I type here in my study on August 2, 2003, I wonder which of my discoveries will have the most important affect upon the lives of all humanity, and if I will be alive to witness some of the changes.

Proving that astrology, that is, certain planetary configurations, has a powerful affect upon human behavior, served as a foundation from whence all other discoveries were derived. In addition, proving that all people are not affected equally from the same astrological vibrations was a challenging project. Then, proving that energy dissolves, the Sun has a gyroscopic rotary gravitational field, and magnetism may bend light on thermonuclear active stars more than gravity must be considered as important.

But the discovery that will make the greatest change in the lifestyle of all humanity is the discovery of how to extract hydrogen from water in huge quantities, in an inexpensive method. This discovery has put a sorrowful burden on my spirit. Where I can find honest people troubles my mind greatly. Certainly, there are huge corporations that would want my knowledge or intellectual property. Yet, as the history of corporate fraud, thievery, and deceit caution me, I cannot simply give my knowledge away to greedy people that would use my intellectual property to make billions and want to hire me as their janitor or gardener, then at times send for me and ask, "What else do you have for our corporation?"

Yet, take heart with me, for I have a plan to give much of the profits from a new hydrogen industry to the American people. My plan is to have 49 percent of the profits go to a Social Security Fund. It is a bold and philanthropic plan that could supply less expensive fuel for all internal combustion engines, and raise the standard of living for all retirees while freeing ourselves from the Arab Oil Cartel. We have to wait and see if those in the federal government will be honest and *not* try to connive a way to give it all to big corporations.

Within my long journey for *My Search for Truth*, I became aware that fellow truth-seekers in the world were eager to know truth about their own nature and the nature of the universe. I am pleased to have served our heavenly Father and his only son, Christ the Lord, to be a vehicle in bringing forth new knowledge that will establish a new physics and a new astronomy. At some time in the future, God only knows when, these truths will be revealed. That energy dissolves is a proven event and this establishes a true understanding about the nature of energy. The law of conservation of energy and the first law of thermodynamics are proven to be opinions and not laws.

Ether is reintroduced as a primal neutrally charged substance (particles) for all electromagnetic forces and matter. I have named the neutrally charged particles *ethons*. I submit that ethons of the ether cannot be destroyed; otherwise, the universe would eventually dissolve. I further submit: these building blocks of all creation acquire polarity and physical properties when they swirl in a vortex.

Electrons are examined and the nature of electrons is submitted. Electrons are composed of a full phase of electromagnetic forces.

Electrons are ejected by an electron emissive material when light is directed on the surface because the light wave is formed into electrons by being captured by the swirling fields in atoms of the substance.

These excess electrons will by ejected due to the atomic unbalanced condition. Nuclear imbalance is also the reason for the nucleus ejecting particles and is known as radioactive decay.

Electricity is explained in this work in great detail from an "in depth" perspective.

Electrons are formed from a phase of electromagnetic energy including light is fact one concerning the nature of electricity. Fact two reveals that light energy and all electromagnetic energy dissolves into the ether when no transformation device is provided.

These two facts given here greatly expand our understanding of electricity. Electricity, being a driven flow of electrons, is universally accepted to be the nature of electricity, but this is where the consensus ends.

When I built my energy-dissolving box, I conferred with an electrical engineer about the number of electrons that were in the current that was transformed to blue light. The engineer explained to me, according to what he was taught in electrical engineering, that no device takes electrons from an electrical circuit excepting a miniscule amount. He said that electrons have mass, and according to Einstein's equation, a tremendous amount of energy is released when electrons in electrical energy is transformed to other types of energy. He continued by saying that electrons are part of atoms and the power company at the generation plant forces them to move through transmission wires and are returned in a loop circuit, excepting a very few, to continue the transmission circuit.

His explanation was totally incorrect. The truth is: our entire electrical and electronic industry has been built on the belief that electricity is a flow of electrons. This is fact, and Ohm Laws in direct current enabled all electrical calculations. However, a very in-depth understanding about the nature of electricity is missing because scientists and engineers still believe the concept given to me by an electrical engineer.

As you read, you will discover the true nature of electricity by in-depth analysis of the factors related to electricity.

Electricity is composed of three components: electrons, electromotive force that accompanies and drives the flow of electrons, and an electromotive field that surrounds a flow of electrons. When the word *electricity* is broken into its syllables, the syllable *tri* denotes three components that have an electric charge and are interwoven in a complex yet orderly interaction.

However, I do not presume to have all the secrets because electrons are also subatomic particles and nuclear science will also have a rebirth as a new physics and a new theoretical physics begins to take shape in the scientific world. No doubt that pions and gluons will be replaced by a more intricate theory that includes the pycnosis concept of matter.

Electrons are not the only particles created by a full phase of electromagnetism for protons and neutrons had an origin.

The original protons and neutrons probably originated from an etheric vortex created by God the Father. After galaxies and stars were created, I submit, in theoretical conjecture, that crushing pressure in the center of a star actually compresses the light wave into a dense condition where the light wave with its photons are formed into protons and neutrons. I further submit: these dense waves are radiated during Sun spot activity and have all the characteristics of the light wave, excepting their great mass. Also, these matter waves could have been formed by the crushing force on other electromagnetic waves, not exclusively to light. Although, if that speculation is true or not true, it means not a thing to we earthlings except to add to our speculations as to how matter was/is formed

Aust is a new name for the gyroscopic, rotary, Newtonian gravitational field that extends from the central girth of all rotating celestial bodies. Therefore, the tubular confines that is created by the Aust holds the planets in place and causes the planets to revolve around the Sun in paths that are uneven ovals, not elliptical paths. Since each planet takes the line of least resistance as it travels in the tubular confine and moves to the outer extremes—up and down and in and out—and the inner extreme in its revolution, this creates an uneven oval shape. Kepler first discovered this and was struck with awe when he saw the pattern.

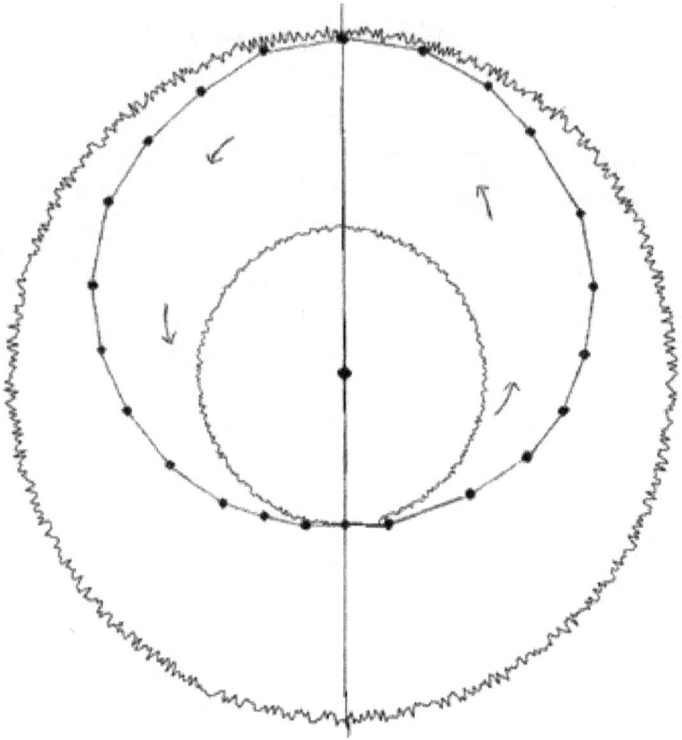

*Tubular confines* is the accurate term to describe the shape within the Aust for every planet and moon's revolving motion around its controlling body. A semi-flattened doughnut shape is a fair description.

Planets move to the outermost position (aphelion) in its tubular confine to the inner most position (perihelion) and this eccentric motion creates or traces an uneven oval in its tubular confine. Planets also move to the upper most extreme of the tubular confine to the lowest extreme.

Magnetism is proven to be a circuit force that exits at the North Pole and enters at the South Pole.

Two claims yet to be proven in a laboratory: In addition to the energy dissolving experiment, two additional tests are submitted to prove the existence of an Aust and to prove that light bending as it passes a star could be caused by a magnetic field instead of being caused by a gravitational field as Einstein claimed in his theory of special relativity.

This synopsis given within the preceding ten claims reveals the basic intent of this work.

Before we continue, it is important for all interested people to know how the gobble-de-gook theory of Einstein's called the general theory of relativity ever came to be so revered by scientists in every advanced nation. In 1905, the bending of light as it passes through an intensely heated atmosphere or an intense magnetic field was known. It was also known that light does not travel at equal velocities in different transmission mediums from the simple observation that the hand seems to bend as it is partially submerged in clear water.

From these known facts about light, Einstein correctly deduced that light would bend as it passes the powerful field forces of a star. The amount of bend should also be determined by the mass of the star. He was also correct in this assumption. Using James Clerk Maxwell's lightwave mathematics with Newton's universal law of gravity, he calculated the correct amount of the light's displacement as it passes a star.

That was in 1905. And in 1915, Einstein hobbled together several concepts that were advanced by geniuses that were working in the realm of physics. He used several very true conditions and perhaps never thought of by theoretical physicists, to weave diverse concepts of others into one gobble-de-gook idea called the general theory of relativity.

The theory probably would have never been given respect

excepting for a tabloid-type news article that was printed by *The Times* of London. In 1919, the Brits had made a second attempt to make comparison photographs of a star to determine if light did indeed bend as it passed a star. The second attempt during an eclipse was successful and *The Times* is chiefly responsible for fanning the science-hungry world into a state of amazement.

After *The Times* printed the headline on November 7, 1919 about a new revolution in science that establishes a new theory of the universe and overthrows Newton's ideas, other newspapers that did not know one iota about what they were printing, in their one-upsmanship, idolized Einstein. From this tabloid-type distorted data, the name Einstein was popularized and a scientific idol was created, even though general relativity is based upon several false ideas. No wonder Einstein said, "Am I or the others crazy?" when he saw this hype developing.

That was November 1919, and now in 2005 the Brits are at it again. If they are hungry, it is time for them to begin to eat their own words. Earlier I illustrated how the BBC was wrong. In the BBC website that was titled Science, Nature And Space, the Brits did not even mention one of the greatest mathematicians and astronomers, Johannes Kepler. His laws put opinions to rest, and gave humanity laws that proved the Sun is the center of the solar system, and the planets revolve by precision movement within the Sun's rotating gravitational force.

Now you have the truth: *The Times* printed a false and misleading article in tabloid style in 1919.

That paper is responsible for beginning the conditioning process to convince the public to believe and revere a false theory. Ah yes, readers, tabloid hype, whether it be falsehoods or blatant lies, convinces a lot of people.

Let us return to the basic intents of this volume. There

have been ten claims submitted in this work, and each must be proven. Also, there are two tests submitted that will prove Einstein's theory of general relativity to be a false theory. If it is true or not true, what difference does it make?

Research scientists know that the propositions in general relativity have no bearing on their research and they know that it has no practical application to building advanced devices or new machines. The Swiss scientists that devised super, super magnets $Sm_2 Co_{17}$ did not rely on Einstein's "warped and deformed space in the vicinity of magnetized matter," for they proceeded with their concept of magnetism rooted in the molecular theory of magnetism, and super, super magnets give positive proof to the molecular theory of magnetism. Also in 1982, General Motors and a Japanese Special Metals Company developed a super, super magnet called a NIB or NEO magnet. This is the strongest magnet yet known. It is made of an alloy of neodymium, iron, and boron ($Nd2Fe14B$), which forms a tetragonal crystalline atomic structure. This crystalline structure causes the electrons to permanently be aligned in a north-south alignment which gives the alloy an extreme magnetic force.

The researchers that developed the NIB magnet also ignored Einstein's idea of a "warped and deformed space in the vicinity of magnetized matter" and proceeded with the molecular theory of magnetism. Even though these two illustrations should be enough to end the debate about the Earth's magnetic force, I offer a test to finally prove conclusively that magnetism is a circuit force.

My postulate concerning magnetism is: magnetism is a circuit force, it flows out of the pole referred to as the North Pole and flows into the opposite pole or South Pole, leaving no room for the existence of a magnetic dipole. If several magnetic atoms could be formed into an atomic molecule and the South Pole could be inverted to be in the center of the molecule, it would give the impression of being a dipole.

It has long been known that the Earth has a protuberance or a pushed out circular mound at the North Pole that is elevated or sticks out fifty miles from the surface of the Earth, and a flattened area at the South Pole pushed in a distance of fifty miles. A polarized iron content of the Earth's heavy metal core could rotate from the gravitational effects of the Moon, Mars, and Jupiter.

Over long periods of time, 100,000 years and longer, these three celestial bodies pulling on the Earth could account for the changing position of the Earth's magnetic poles, and the magnetic flow would account for the new poles to cause the pushed out and pushed in shape upon the Earth's polar regions. Earth's inertialized magnetic iron core cannot move unless it is acted upon by a force. The power of the Earth's three greatest gravity-exerting bodies (as given) is that force.

During 1966, NASA officials revealed accumulated data from the lunar orbiter that revealed a quarter of a mile bulge at the Moon's north pole and a one-quarter mile depression at its south pole. I submit that these deviations in the Earth's and Moon's contour are caused by a circuit of magnetic force. It has been said that a picture is worth a thousand words. The following picture conveys great truth in graphic form. The picture depicts the north magnetic pole to swirl in counterclockwise motion and the south magnetic pole to swirl in clockwise motion.

I admit—I do not know which is clock and which is counterclockwise. The direction of the swirling motion is a project for other researchers.

## EARTH

### Counter clockwise motions

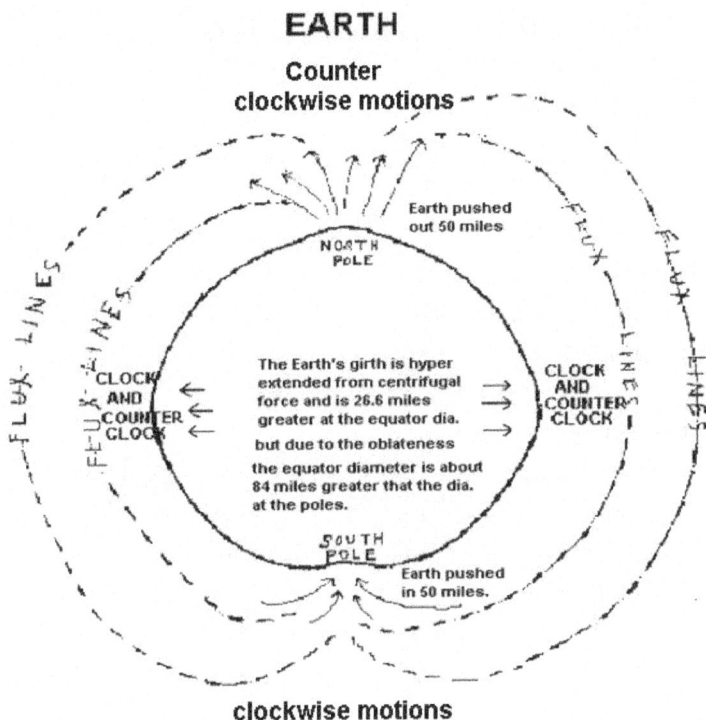

FLUX LINES

Earth pushed out 50 miles

NORTH POLE

CLOCK AND COUNTER CLOCK

The Earth's girth is hyper extended from centrifugal force and is 26.6 miles greater at the equator dia.

but due to the oblateness

the equator diameter is about 84 miles greater that the dia. at the poles.

CLOCK AND COUNTER CLOCK

FLUX LINES

SOUTH POLE

Earth pushed in 50 miles.

### clockwise motions

Remember, light bends as it passes a star.

Question 1: Why does light bend as it passes a star?

Question 2: Is the bending property equal over the entire surface of the Sun? The answer to that question is very important.

Question 3: Does gravity and/or magnetic force slightly deflect light as light passes through a powerful force field?

Let us first consider the possibility that gravity can bend light. If gravity can bend light, I submit to the entire body of knowledge in physics and astronomy a new concept.

I submit: If gravity can bend light, the greatest refractory power is at the region of its greatest girth—or—at the Aust closest to the rotating Sun. This idea can be proven by a test in a laboratory.

Gentlemen Of Science: Don't Scorn this Idea. Conduct the Test. Remember the Tests Performed By Cavendish About 350 Years Ago.

The test that I submit requires a narrow beam of light (preferable a blue light, since it will bend more easily compared to the other colors or the full spectrum) to be focused on a heavy spinning spheroid.

If gravity does bend light, it should bend at the bulged area if the mass of the spheroid is sufficient.

Does the magnetism of a sun bend light as light passes?

A careful test beyond Lorentz and Zeeman is needed.

If the North Pole of a magnet has a force flowing outward, simply by making a sphere from super, super magnet (NIB) material [and perhaps cryogenic] and focusing light through a small opening on cardboard that covers the sphere then through different places at the North and South Pole, light should be able to be deflected or attracted. Blue or violet light should be used because light at the upper end of the spectrum has a greater bending characteristic. However, don't be restricted to blue light; use all the colors and the full spectrum, or clear light, and discover the laws given by James Clerk Maxwell and Isaac Newton have not been invalidated.

Magnetic force from the north pole should push the light away from the sphere, while the south pole should pull the light - toward the sphere.

## AN AUST INHERENT TO ALL SPINNING SPHERICAL BODIES CAN BE PROVEN

To prove the existence of an Aust in a laboratory: direct a blue light downward upon the spinning magnetic gyro top or a sphere to determine if concentric circles of more intense blue light will appear at the region outward from the spinning spheres greatest girth. Alter the rotation movement faster and slower to determine the effect. If the researchers performing this test still want to hang on to the theory of general relativity and the inertia idea, the results of this test should impart unto them greater perception.

If my expectations would prove to be correct, two natural phenomena will be revealed:

First, magnetism can again acknowledged the molecular theory of magnetism, that is, magnetism as an inherent force within atoms. Therefore, magnetism is a property of mass—and not a "condition of space in the vicinity of magnetic matter." Also, and the bending of light as it passes a star can be re-evaluated as to the reason. I certainly believe that gravity can bend light. At the Aust of a rotating sun and as light approaches a "black hole" or a new "neutron star"?

Secondly: the discovery of concentric circles of blue light as given should supply impetus to accepting a rotary force field, thus an Aust.

The truth about one of Einstein's speculations is the fact that he was knocking on the door of understanding when he postulated the existence of a "warped and deformed space." However, his reliance on the false ideas of other theoreticians prevented him from grasping the truth about a rotary gravitational field. When Einstein rejected the Ether, rejected the pycnotic concept of matter, and relied on his idea of inertia for planetary movement, he created a trap for himself where there was no escape. In addition, the absence of facts to support his flawed ideas would eventually doom his theory.

The phrase "a rotary gravitational field" will become common. These phenomena with proof that energy dissolves will shatter the theoretical ideas postulated in general relativity, re-establish the Kepler, Newton-Cavendish laws of planetary motion and enable the Aust to be accepted into the science of Solar Dynamics. Below is a graphic of the device that will prove the existence of the Aust. A great era for physics is on the horizon.

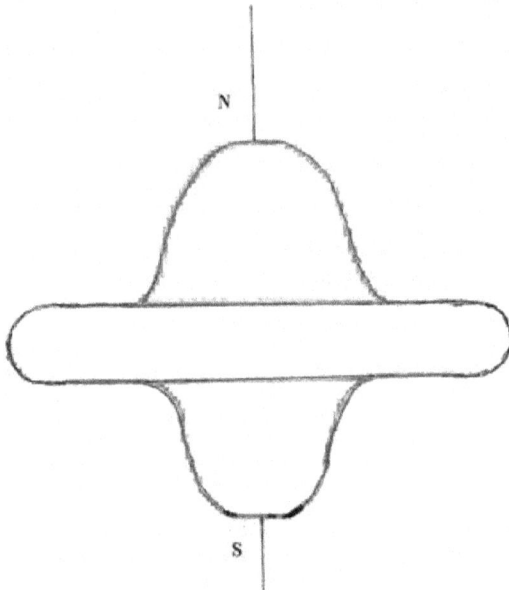

It has been over twenty years since I wrote this manuscript and in the time since the early 1990's scientific minded men have brought forth new discoveries. One outstanding discovery that proves the existence of -- the Aust - has measured gravitational anomalies over the entire earth. A renowned physicist - Dr. Alan Ward - posted on the Internet dated - March 20, 2004 - graphics that prove the existence of the Aust. Although men in astro-physics know of the anomalies but no one except this metaphysician knows what the anomaly pattern means.

It has been said that a picture is worth a thousand words. See for yourself the actual gravitational pattern, but first read the quote from Sr. Alan Ward.

"The Gravity Recovery and Climate Experiment (GRACE) is Revealing More Detail About the Gravity Field than has been Available Before."

The Aust band is clearly shown where water does not swirl as it goes down a drain, and the region above and below the Aust where the fast-moving gravity force meets a slower moving gravity force and the two distinctly different gravity forces interact with each other and swirling gravitational forces are formed that I have named the Wake. One swirls counterclockwise and the other clockwise, above and below the Aust, at the Equator.

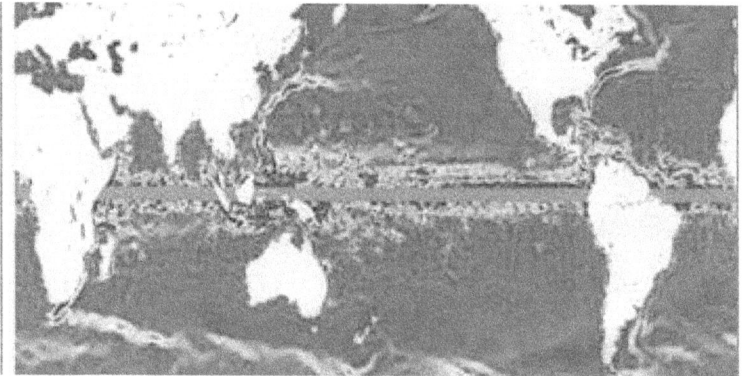

We Continue:

Ptolemy (about 100 AD–170) and Einstein (1879–1955) stand out as giants in the annals of scientific history for being the greatest at promulgating false concepts. It is true that they did not take the reins in hand that chart our development course, but they were the two most outstanding leaders in the scientific world of theoretical science that were successful at convincing influential leaders to believe in and accept their false theories. As stated previously, it was the controlling clique of enthralled supporters that held the power to decide what is truth and what is not truth that are to blame for the long periods of teaching false theories. If it wasn't for *The Times* hyping general relativity, it probably would not have got off the ground. Although truth does eventually triumph and Ptolemy's erroneous concepts based upon his Earth-centered universe idea was eventually replaced by a true understanding by Copernicus and Kepler.

We would be at fault if we would expect every analytical scientist during several thousand years of scientific development to be correct when grappling with the secrets of the universe. Therefore, a degree of magnanimity is necessary to peer into the reasons for any mistakes in a science so complex and profound as theoretical physics, especially when considering the constant condition of our limited understanding.

Therefore, it is logical to expect a degree of error in a science where the propositions are not based on actual data from tests. Yet as you read you will find that my criticism of Ptolemy, Einstein, and organized science in the last half of the twentieth century is often scathing, and so it is with debates. Each side must be ready to defend their position and be responsible when submitting data to give credibility to their propositions. But, as Shakespeare said, "all the men and women merely players," which moves me to pay honor and respect to every scientist, those who touted false theories and those who revealed truth.

**A JOURNEY INTO THEORETICAL PHYSICS BEGINS.**

It is now time for us to begin a journey into theoretical physics.

I know that once you have embarked on this journey you will agree that indeed an impressive journey has begun. So find a quiet place, and I will take you on a journey into the realm of the abstract and into the solar system without even leaving your chair. Before you put on your thinking cap, I wish to share with you a philosophical verse that I wrote in the early 1960s. It is as follows: "New Theories Rise and Great Theories Fall but Truth Remains, Waiting for Man to Call." So truth will greet you as you read and become aware of the falling of great theories. However, there is much to learn.

Before we begin that journey into theoretical physics, let us return again to a time when I was very young. I had the good fortune of knowing some very brilliant and super talented people, beginning when I was young. I have known geniuses in the realm of music, art, mathematics, and other mental wizardry.

The wizardry of these individuals emerged when they were very young. Although being very gifted, in other ways they were normal and no one could have guessed or ascertained their super ability. My father was a wizard at the piano keyboard and may have followed music if his interest in science had not taken precedence. If he didn't know how something worked, he would take it apart to discover how it worked. My first son also has this trait. About the year 1930, my father had founded a radio and sound system business. My mother died that year and my father was devastated. Six years later, my father died of a heart attack.

Being near 80 years old, (as I review these manuscripts I am near 85 years old) I have wondered about my scientific acumen. The best explanation that I have is that it comes from three sources: from my German lineage, from the law of compensation, since I am slightly incapacitated from the effects of osteomyelitis, and from a psychic source.

The third source I attribute to God the Father and Christ the Lord. I have a psychic ability that operates intermittently. Sometimes I know what people are thinking. This psychic ability has become stronger since 1958 when I first began to have out of the body experiences. The greatest secrets of the solar system have come directly from higher consciousness.

The Aust, with the sixteen-point orb structure of each planet, I know for sure came from higher consciousness. Twice I was concentrating on a solar system problem and I suddenly wheeled my chair backwards away from my desk and I felt a slight giddy feeling, while at the same time it seemed as though I was coming through a tunnel. As I was through the tunnel, the giddy feeling vanished and I felt normal. I felt similar feeling when I came out of my body, so I know the information was from a higher consciousness. Since I often pray to God our Father and Christ, I attribute some of the knowledge to have come from either source. All that I can say about being different from others is the fact that my zeal to understand the nature of our universe is an unquenchable urge. I have been questioning and analyzing as long as I can remember, and my faith in God and Christ has increased as I became older.

My interest in the solar system began in my fifth year in primary or grade school. I drew a picture of the solar system with all nine planets. The picture hung on my bedroom wall for several years. As the years passed, my scientific research became more intense. Thus, my interest in the solar system led me to study what is known and what other researchers have come to believe from their theoretical analysis. Early in my studies I had learned a very vital fact about the motion of the Sun and the planets. This very vital fact is: the Sun rotates in a counterclockwise direction and all the planets revolve in the same direction, that is, a counter clockwise direction.

Secondly, I discovered that in more than three quarters of the twentieth century, no serious attention has been devoted to this aspect of the solar system dynamics. I believe that other theoretical thinkers would have recognized that a spinning celestial body extends a powerful gyroscopic gravitational field, if they had not been forced to accept Einstein's inertia concept.

Kepler (1571–1630) had suspected the Sun to be the root force that propelled the planets in their orbits and included this concept in his 1596 published work , meaning "Cosmographic Mystery."

However, when I began to seriously study the solar system in the late 1950s I had not studied Kepler's work and I had arrived at the same conclusion as Kepler through my own analysis. That was over 350 years after Kepler's first theoretical work.

By 1958, when I became focused in my study of the solar system, Einstein's ideas concerning solar system motion were greatly respected and revered as the gospel truth. This condition forced me to study Einstein's concepts and to review history to learn how our present understanding of the solar system developed from the work of previous researchers. Thus my sharp focus in mysteries of the universe and my spiritual nature became anchor points that positioned me for expanding studies. Hence, this is how this work put together in a three volume work titled; *My Search for Truth* had a simple beginning.

My cross-examination and analysis of the work of the great alchemist and theoretician Newton (1642–1727) led me to research how and why Newton discovered the universal law of gravity. According to Kepler's second law, a sun force caused planets to move more swiftly as they become closer to the Sun and move slower as they move farther away from the Sun. In addition, all of the planets moved in clockwork movement, since Kepler's third law revealed their distance and period of revolution to be in a precise mathematical relationship, i.e. d3 = p2.

Newton's genius in mathematics perceived a diminishing gravitational force that is determined by the distance that separates mass. But his famous law was only a theoretical concept because the formula lacked the necessary constant. It was the very brilliant and shyest of all scientists Henry Cavendish (1731–1810) that discovered the constant by working with lead balls.

In 1798 Cavendish began to experiment with lead balls suspended by a wire. He measured the amount of twist in the wires as he brought the balls very close. Thus, he used the data to calculate the mass and density of the Earth. It was not a simple task, and it is a work that towers very high in the annals of scientific development, for with this knowledge, Newton's universal law of gravitation became a mathematical tool to calculate motion that is subject to gravitational force. Since my searches found no fault in the Newton-Cavendish law and did find precision accuracy, my research eventually led to Einstein (1879–1955). This eventuality was a natural sequence because I found a general agreement with my basic concept of the solar system and the work of other scientists, but with Einstein's ideas I found not only major differences but irreconcilable differences.

The fact that great scientists, including the Nobel Prize committee, never accepted Einstein's theory of General Relativity strengthened my concept of a Sun-powered rotary gravity field, instead of Einstein's idea of inertia being the force that drove the planets around the Sun.

Einstein had, as we say, died in 1955, and that was three years before I seriously renewed my in-depth scrutiny into mysteries of the universe. So, I never had the pleasure of conferring with him and challenging some of his concepts in a two-way conversation. If we had met to compare concepts, I would have told him the following: "There is *no* factual basis to attribute the revolving motion of the planets to inertia; therefore, please consider an alternate concept that I favor. The Sun's rotating motion creates a rotary gravity field—that is, the Sun rotates counterclockwise, and since it has over 98 percent of the solar systems mass, it controls the gravity field of the solar system. Therefore, this rotary gravity field extends outward from the Sun's greatest girth or its zero line of latitude as a gyroscopic force field. This revolving force field from the rotating Sun is the most logical, based upon certain known facts, to be the driving force that drives the planets and holds the planets in their orbits.

I would have also said the gravitation of the rotating Sun causes the planets to revolve in a counterclockwise direction and controls the speed of their orbital motion according to their distance from the Sun. I'm sure with my injection of counter logic Einstein would have been unable to defend his logic.

Yet in firm truth, unless a meeting between myself and Einstein would have occurred in the presence of witnesses, I very probably would have been very disappointed at the outcome. I base this strong probability that I present on the basis of the following actual occurrence. I recall a national news event that featured Einstein about a year before his demise. This news was of a scientific nature concerning a new perspective of his concept of mass, time and motion in the universe. Einstein had lived in Princeton, New Jersey, since he worked at the university's Association for Advanced Study. Other theoretical scientists also worked there and that particular event that became a national news story originated from a dinner engagement Einstein had with a scientific colleague. According to the fellow scientific colleague of Einstein's, during the dinner he revealed to Einstein a concept of his own concerning mass and motion in the universe. Shortly after that dinner and the theoretical hypothesis that was explained to Einstein, an astonishing news release revealed that Einstein had developed an extension to his general theory of relativity that established a link to his developing unified field theory.

The fellow scientist was incensed over what he claimed was a case of purloined material and lamented over what he called a devious act. In defense of his charges, he publicly proclaimed the new work was actually his and not Einstein's. To my great sorrow, I did not clip that news article, but that story became etched in my mind and I later discovered that purloining material was a normal procedure of Albert Einstein's. While Einstein was weak in probing into mysteries with psychoanalytic skill, Einstein did have great mathematical acumen, but his flawed concepts of physical phenomena or physical laws and his inability to grapple with complex philosophical systems led him into limited understanding and false concepts concerning our physical universe. This is very apparent in Einstein's inability to psychologize in the realm of the abstract, for it led him to give up trying to understand an ordered universe created by a divine invisible mind. Einstein was a disbeliever in any divine creator or a God.

His belief in deity vacillated between his non-belief and references to God. In reference to Heisenberg (1901–1976) and his uncertainty principle, Einstein did make a remark with reference to God, whether it be to strengthen his concept or not we do not know. It is recorded that he said, "I don't believe God plays dice with the world," but when hearing the Austrian born violin genius Fritz Kreisler (1875–1962) play the violin he said to Kreisler, "You have restored my faith in God."

So it depends upon what time or period of Einstein's life a focus is made to determine if he was believing in a God of the universe or disbelieving at that particular time. His lack of ability to understand the greater meaning of life was demonstrated by his lack of firm conviction concerning the nature of a Supreme Mind of the universe. His atheistic view was not a secret.

I have read extensively about Einstein's life and his theories and my knowledge does not come from superficial skimming. In 2008, an Einstein letter written to a friend was sold at Bloomsbury Auctions, in London, for $404,000 (that's four tenths of a million dollars), and in November 2012 it was auctioned again and this time it went for just over $3 million dollars. In the letter, written to philosopher Eric Gutkind, in 1954, one year before Einstein's demise, Einstein stated his disdain for religion in his letter to his friend. Following is a partial quote from that letter.

> The word God is for me nothing more than the expression and product of human weaknesses, the Bible a collection of honorable, but still primitive legends which are nevertheless pretty childish. No interpretation no matter how subtle can (for me) change this. These subtilised interpretations are highly manifold according to their nature and have almost nothing to do with the original text. For me the Jewish religion like all other religions is an incarnation of the most childish superstitions. And the Jewish people to whom I gladly belong and with whose mentality I have a deep affinity have no different quality for me than all other people. As far as my experience goes, they are also no better than other human groups, although they are protected from the worst cancers by a lack of power. Otherwise I cannot see anything 'chosen' about them."

Ugh—and **Golda Meir**, born Golda Mabovitz (May 3, 1898–December 8, 1978) wanted Einstein to be president of Israel. The atheist ignoramus wasn't even qualified to be a good Jew. For all the Einstein aficionados, I challenge all of you to give an in-depth analysis for the creator of the false theory general relativity and his atheistic ideas.

Since Einstein was Jewish, he probably never studied the New Testament. Also, by his writings it is evident that he never studied anthropology, psychology, or comparative religions and therefore never studied the nature of God and the mind of man, as revealed in these disciplines.

In reference to his general theory of relativity, Einstein did say "no amount of experimentation can ever prove me correct but a single experiment at any time can prove me wrong." Read that quote again all scholars of truth, for the tests that I submit are more than "a single experiment at any time can prove me wrong." For the truthful results of three tests: energy does dissolve by the mirrored box experiment, and the rotating super super magnetic sphere will prove that magnetism can bend light. In addition, the existence of an extended "plane polarized" gyroscopic force field, which I have named the Aust, will be proven. The nature of the Aust will be explained as you read.

Therefore, in recognition to this claim of Einstein, that a single test can prove him wrong, weighed heavy in my mind. For a researcher as myself that looks very deeply into the subject material, and at the same time conducts a parallel psychoanalytic study, this is an important claim, for the proponent of radical concepts as Einstein's should introduce ideas where spin-offs should lead to new inventions. This positively is not the case.

In fact, general relativity is a sterile theory because general relativity is a false theory. Einstein's revered theory never led to any invention. Because of this, in the near 40-year span from 1915 when general relativity was introduced to the demise of Albert Einstein in 1955, a Nobel Prize was never awarded for general relativity, for members of the review board that determines the worth of scientific contributions are not easily misled. Einstein's fallacious theory had the effect of stifling development. If this is shocking to you, then please continue to read these pages very thoughtfully.

Many people in and out of science crave to create an idol with such stature that the idol can be idolized, that is, a person can be revered with blind devotion. Some of these chauvinistic devotees claim that Einstein was a great inventor and this claim too is a fallacy. This claim is based on the seven-year collaboration of Dr. Leo Szilard and Einstein. During this time, Dr. Szilard, the super genius, did take Einstein into a development problem involving making a refrigerator with a magnetic force to create rapid heat reduction and thus serve as a chiller in a refrigerator. After Freon was developed, this idea was shelved. Although, the idea is not dead and new discoveries may again revive this idea. But the point I wish to elucidate concerning the many ideas of Dr. Leo Szilard is the fact that he did not need Einstein, and Einstein probably added little or nothing to the ideas of this magnetic-powered refrigeration unit. This idea was never developed into a working contrivance or device. This presents a mystery to me, because ideas are not patentable; only workable devices are patentable. Therefore, I ponder, did the world reputation of the relativity man contribute to patents being issued for a device that never worked?

Leo Szilard had a high respect for Albert Einstein since the London newspapers had set Einstein up as an idol in 1919, and in 1921, Einstein had become one of Szilard's physics professors, along with Max Planck and Max von Laoe, while Szilard was studying at the Berlin Institute of Technology. This association and Einstein's idol status probably were the factors that influenced Leo Szilard to entreat Einstein to work with him in the refrigeration venture. However, this theoretical device never was developed to work, and if Einstein was a great inventor, then he should have brought forth some inventions in his lifetime on his own, without a collaborator, but he did not.

Consider the contributions of Dr. Leo Szilard, the real genius and father of the atomic bomb. In 1923, Dr. Szilard collaborated with Dr. Hermann Mark on X-ray diffraction experiments, in Berlin at the Kaiser-Wilhelm Institutes for Chemistry. In 1926, Dr. Szilard began seven-year collaboration with Albert Einstein on a home refrigerator without moving parts. In 1929, Dr. Szilard filed a German patent application on the cyclotron. In 1931, Dr. Szilard filed for a German patent on the electron microscope. In September 1933, Szilard conceived the basic principle of the atomic bomb. In 1934, on March 12, he filed for a British patent on the atomic neutron chain reaction.

In 1934, Dr. Szilard, with the collaboration of a fellow scientist Dr. Chambers, invented the Szilard-Chalmers reaction. It was a method for concentrating artificially produced radioactive isotopes. In 1937, Dr. Szilard, working in collaboration with Dr. James Tuck, designed a betatron. In 1942, Dr. Szilard designed cooling systems for atomic reactors.

In 1944, Dr. Szilard proposed the term *breeder* to a reactor that was able to generate more fuel than it consumed. In 1948, Dr. Szilard invented the chemostat, an apparatus to enable a continuous production of bacterial cultures. In 1955, Dr. Szilard, in collaboration with Enrico Fermi, was issued a patent on a nuclear reactor.

This list does not include the total contributions made by Dr. Leo Szilard. Although, this list should be sufficient to illustrate the actual genius mind working for forty-plus years in several scientific disciplines—and the lack of contributions brought forth by Einstein.

There is a widely held belief that Einstein had a working position with the development of the atomic bomb. This idea is not true, and to clarify any misconceptions, I have included a history of the development and final use of the atomic bomb.

**THE HISTORY OF THE DEVELOPMENT OF THE ATOMIC BOMB.**
At humanity's present state of development, credit to the wizardry that comes from the human mind is erroneously attributed to the ego consciousness. However, in accordance with the present-day understanding of the human mind and to avoid any metaphysical futuristic concepts, I will refer to the ego identity of each individual that participated in the development of the atomic bomb.

For nearly 2,000 years the understanding of the basic building blocks of substance (atoms) rested with the theory of Democritus (460? –380? BC). The theory that has come down through the centuries from Democritus pertains to the smallest particle of matter. According to his theory, matter can be cut into smaller and smaller pieces until further division is impossible. This smallest particle of a substance while still retaining its characteristics was called an atom for the word *atom* is a Greek word meaning "non-cuttable," and this was derived from the Greek word *tomas,* meaning "to cut."

Then about 2,000 years after Democritus, during the 19th century, instruments were devised that enabled research into the nature of the atom, and three researchers—Gustav Robert Kirchoff (1824–1887), Johann Jakob Balmer (1825–1898) and Ernest Rutherford , 1st Baron Rutherford of Nelson (1871–1937) —became the three outstanding early researchers that laid the ground work for a more comprehensive understanding about the nature of the atom.

Rutherford was the first to speculate that atoms had a center (which he named the *proton*) and is surrounded by a cloud of electrons. In 1885, Balmer released his discovery about successive steps of atomic spectra, which later became famously known as "The Ladder of Balmer." Eighteen eighty-five not only stands out in importance as the year that Balmer released his formula, for in 1885 a Danish baby was born of a Danish physiology professor and a Jewish mother who was to become a world famous physicist.

That future world-renowned scientist grew to become the famous Danish physicist, Niels Hendrik David Bohr, (1885–1962.) Bohr extended our understanding of the atom one more step beyond the theoretical work of Rutherford, and in doing so, he became the founder and therefore the leader of our modern understanding about the structure of the atom. For decades, scores of years, and centuries of time to come, the Rutherford-Bohr concept of the atom has and will endure. However, the understanding about the cohesive force that binds the nucleus together has eluded scientists.

This scientist (myself) firmly believes the pycnosis concept of matter is true and also believes when scientists, in error, rejected the pycnosis concept, they moved farther away for understanding atomic cohesive force. Yet, in time, truth will triumph. But before science would focus on the root attachments of atoms to the mother particles of all creation, ethons of the ether, the energy released from the fission and fusion of atoms would take the forefront due to political conditions and warfare technology.

As we trace the development of the atomic bomb and the scientists that were involved in bringing it to fruition, we move forward to the analytical rich mind of Niels Bohr. He received his doctorate at the University of Copenhagen in 1911 and shortly afterwards he went to Cambridge in England to further advance his knowledge.

While Bohr was in England he studied under the famous physicist Sir Joseph John Thomson, (1856–1940) whose work with cathode rays offered conclusive proof that cathode rays were composed of subatomic particles and this proof placed another building block in understanding the nature of the atom.

Later, after Bohr moved to Manchester, he studied under the Scottish physicist Ernest Rutherford. Rutherford's speculation about an atom's central core or nucleus being composed of a proton and being surrounded by a cloud of electrons made a strong impression upon the mind of Bohr for he related this possibility to the work of previous researchers, especially the work of Balmer and Max Karl Planck, (1848–1947) and Ludwig Ernst (1858–1947). Bohr theorized that atoms could absorb and emit energies in quantum amounts by reason of the fact that an atomic cloud composed of atomic energy and electrons were absorbing and emitting quantas of energy. He correctly reasoned that this possibility would explain the work of Fraunhofer and a century later by Kirchoff, but when he considered an atomic order that was advanced by Balmer, he conceived an order in atomic workings that catapulted physics into the nuclear age.

In 1913, while Dr. Bohr was 28 years old, he cut a pioneering path for all following physicists by releasing his theoretical concepts about nuclear physics that explained an atom's structure as a miniature solar system. His work postulated the center of an atom, as the Sun in our solar system, has the greatest mass and electrons, as the Sun's planets, revolved about the center or nucleus of the atom.

The value of his work proved to be absolutely essential in furthering the development of nuclear physics, and nine years after releasing his theory, in 1922, he was awarded a Nobel Prize for his outstanding work. After completing his studies in England, he had moved back to Denmark and joined the faculty at the University of Copenhagen and later helped to found the Institute for Theoretical Physics at the University.

Dr. Bohr came to America in 1938 and early 1939 to work at the Institute for Advanced Study at Princeton, New Jersey in research to determine crucial facts about Uranium 235 (U-235). When Hitler's army invaded Denmark in 1940, Bohr ordered all work at the Institute for Theoretical Physics stopped. As Hitler became a greater menace, Dr. Bohr's life was threatened, since Bohr was 50 percent Jewish, but before the Nazi henchmen could capture Dr. Bohr, the British helped Dr. Bohr escape and took him to England. Later in 1943, he came to America where he became an advisor to atomic bomb scientists at Los Alamos New Mexico.

During my research about the history of the development of the atomic bomb, I found one picture that was taken at Los Alamos for history in the CD titled *Critical Mass* (Corbis release) of Niels Bohr, Robert Oppenheimer, Richard Feynman, and Enrico Fermi standing together with a news paper that blazoned the headlines about the spy network that enabled the Soviet Union to build an atomic bomb.

Through Klaus Fuchs (a minor scientist turned spy that

worked at Los Alamos) Harry Gold, David Greenglass, Julius and Ethel Rosenberg, and Russian spies in America, the vital information was sent to Russia where the Jewish-Russian scientist by the name of Andrei Sakharov headed the copycat Russian team that resulted in the building of an atomic bomb, even down to the one mistake in the wiring.

It is a tragedy that after brilliant Jewish scientists conceived and worked as leading scientists in the development of the atomic bomb that four brilliant Jewish people turned to being a spy against America, but it is a clear illustration that being brilliant does not immunize anyone against mental affliction.

The four Jewish spies were definitely afflicted with a schizophrenic obsession if not possession.

*A side note pertaining to Russian copycat methods is the exact copy of the American B-29 long-range bomber. In the late 1940s, Stalin gave his scientists and engineers two years to build an exact B-29 duplicate. This strenuous effort actually molded the Russian airplane industry. They subsequently built over two hundred copycat B-29 planes. Following the copycat B-29 plane, more success came when the Russians built the MIG airplane. It would have never been developed if the Russians had not bought a completed, operating Rolls-Royce aircraft engine and copied it to the finest detail, manufactured it and put it in the aluminum plane named the MIG. Also, after the Western geniuses became aware of the Russian copycat tactics, plans for the French supersonic Concorde were altered to include a flaw and left at a convenient place for a spy to steal. The Russians then made a copycat supersonic plane that crashed, and they did not try to make another since word was leaked that they had built a plane with a built-in design flaw.*

**RETURNING TO NIELS BOHR**

Following Bohr's release of his monumental work in 1913, scientists in Europe and America began speculating that a tremendous amount of energy could be unleashed if atoms could be caused to have a chain reaction where they would be split into a lighter atomic identity. Then in 1915, Albert Einstein had incorporated into his theory of general relativity his famous formula (which is a variation of a formula of the renowned French physicist Coriolis) that theoretically conjectured the quanta of energy that would be released if atoms could be split.

While the ratio of the amount of energy released to the amount of mass used is highly speculative and is impossible to prove, the formula did impress some theoreticians. But the development of atomic energy was not contingent upon Einstein's famous formula, for scientists believed a tremendous quantity of energy could be available whether it was less or more than the mass multiplied by squaring the velocity of light, if atoms could be split.

But, Einstein's formula did bolster the faith of some nuclear physicists that were speculating about the means to cause a continuous nuclear reaction or fission. Although some German scientists (in fact, the development of nuclear science came from chiefly German and German Jewish scientists) that were on the front edge of nuclear science did not accept Einstein's laced-together theory of other theoreticians which Einstein named general relativity for not all German scientists accepted Einstein's formula, in truth many German scientists completely rejected his theory of general relativity on the basis of what they believed to be preposterous or wild speculations.

So even if Einstein had never been born, research in nuclear science was advancing during the decade of the twenties. At this time in Germany, when nuclear physics was on the mind of brilliant theoretical physicists, the scientist that was to become the leader in developing the atomic bomb was teaching at the University of Berlin in Germany. His now-famous name is Leo Szilard, (1898–1964); he was born in Budapest, Hungary of Jewish parents as was Edward Teller, (1908–2003) and Eugene Paul Wigner, (1902–1995.)

Szilard was a super genius and he stands out as the real father of the atomic bomb as Teller stands out as the father of the hydrogen bomb. For no other scientist could be given the greatest credit for setting the early focus into the fission process that led to the development of the atomic bomb, than Dr. Leo Szilard.

It was no accident that the young Leo Szilard was motivated towards science, for his father was an engineer and speculating about the means to control and capitalize upon the physical environment was certainly part of his early conditioning. His interest in studying the nature of the physical environment led him to study physics in Germany at the University of Berlin.

In 1922 at the age of 24, Leo Szilard obtained his doctorate at the University of Berlin and subsequently joined the faculty. As soon as Hitler came to power in 1933, Dr. Szilard left Germany, one day before Hitler had the emigration escape route closed.

In 1932, a further advancement in nuclear science was achieved when James Chadwick of England discovered the neutron. This discovery was enabled by the findings of two researchers, one in Germany, Walter Bothe, and Frederic Joliot-Curie in France. This was an important step in nuclear science, and in 1934, experiments in Rome that involved Enrico Fermi, proved that neutrons could be used to transmute atoms including uranium.

News of this discovery spread in Europe and Dr. Szilard's interest in nuclear science grew. It didn't take him long to perceive the basic concept of an atomic bomb. In 1934, Dr. Szilard conceived the principle of a chain reaction by using neutrons to cause an atomic breakdown or atomic splitting, whereas two neutrons would be released and break down or split two more atoms and a chain reaction would vastly multiply resulting in an explosion. Szilard perceived the enormous military use for an atomic bomb and applied for an English patent on the process while keeping his idea secret and influencing the British to keep it secret, for even in the early 1930's Szilard was suspicious of Hitler's intentions.

However, while Szilard was correct in his perception as to how to cause a chain reaction that would result in an atomic explosion, he envisioned the use of beryllium to be broken down into helium, and this method was not possible, for atoms of these two elements would not fission in a continuous chain reaction. Although Dr. Szilard did not know the elements that he envisioned to cause a chain reaction were not fissionable material, he was convinced that a fission chain reaction could be engineered to cause an atomic explosion.

While Szilard was working in England, events were also unfolding in Germany, and a Jewish woman physicist that had come from Austria became directly involved in research with a German physicist by the name of Otto Hahn, (1879–1968).

The now renowned female physicist by the name of Lise Meitner, (1878–1968) (MITE-ner) was born of Jewish parents in Vienna, Austria. She studied at the University of Vienna and obtained her doctorate in 1906. In 1907, she went to Berlin to attend lectures of Max Planck and stayed to work with Otto Hahn as a duo team member in scientific research that lasted thirty years. Dr. Lise Meitner was no stranger to Dr. Leo Szilard, for in 1930 he had taught a seminar on chemistry and nuclear physics with Lise Meitner.

In 1936, the duo of Hahn and Meitner, discovered that uranium would successfully fission when bombarded by neutrons, although the discovery was not revealed in a publication. But only two years later in 1938, another Jewish-German physicist by the name of Hans Albrecht Bethe, (1906–2005 ) described the fusion process that powers all suns or stars.

During the decade of the thirties, the interest in fission and fusion by German scientists was becoming more focused and concentrated. During this period, the German Nobel Prize-winning physicist Werner Karl Heisenberg, (1901–1976) was conducting theoretical work on fission and eventually became the head of Hitler's atomic bomb program, although Heisenberg was against building the atomic bomb for moralistic reasons and Hitler's program while active had never reached a fever pitch.

Although, as the new science of nuclear physics was emerging in the 1930s, political events were again changing in Germany, and in 1938 Dr. Lise Meitner successfully escaped Germany to escape Hitler's henchmen and went to Stockholm, Sweden. After her departure, the position she held with Otto Hahn was filled by a chemist by the name of Fritz Straussmann (1902–1980 ), who had received his education at the Technical Institute at Hanover. Neither Hahn nor Straussmann were sympathetic to Hitler but remained quiet about their political inclinations and kept their nuclear research secret. These men worked out an understanding about uranium fission and Dr. Meitner, while in Sweden, was kept informed about Hahn's and Straussmann's work.

Also further south in Italy, Enrico Fermi (1901–1954)

the Italian physicist whose seats of tenure were at Rome and Florence, was also working with fission. In 1938, Fermi received a Nobel Prize for his work in nuclear science. He and his family went to Sweden to receive the prize, and while there the Italian Fascist press criticized him for not giving a Fascist salute and not wearing a Fascist uniform. He was married to a Jewish woman and for reasons of wanting safety, security and freedom he never returned to Italy. He then journeyed to London and from there to America, which also afforded him the environment to continue his research with fission.

He secured a teaching position at the University of Chicago where Dr. Leo Szilard and Dr. Eugene Wigner, (1902-1995) were on the physics staff of the Metallurgical Lab. While the Europeans were working with constrained vigor on nuclear physics, so also were American physicists engaged in similar research. *World Book* states, "Alfred A. O. Nier first separated the two isotopes of uranium, U-235 and U-238 at the University of Minnesota." Pure U-235 was caused to have a chain reaction and therefore a mini-explosion was achieved by John R. Dunning at Columbia University in January, 1940.

During the late period of the 1930s, without government help, American physicists were working independently in universities to harness the power of the atom by fission. Then in 1941, a group of scientists, including Enrico Fermi, Leo Szilard, and Eugene Wigner, working under Arthur H. Compton at the University of Chicago, sustained a continuous nuclear chain reaction.

When the Manhattan Project was begun in 1942, Arthur H. Compton, Enrico Fermi, Leo Szilard, and Eugene Wigner went to work with atomic scientists at the White Sands Los Alamos atomic bomb research facility in New Mexico. As we review the history, it is evident that 1939 was a pivotal year that catapulted nuclear research in university laboratories to the American Manhattan Project. The shift came by the simple act of Dr. Niels Bohr coming to America to attend a scientific conference. However, he also brought news that Dr. Meitner was soon to publish the discoveries made in Germany about a chain reaction of fission occurring with uranium when it was bombarded with neutrons.

Dr. Lise Meitner had a scientist nephew that was a British citizen by the name of Otto Robert Frisch, and through them the scientific world was informed that a process of fission or a chain reaction was successful in Germany. Dr. Meitner published her first report in January 1939. The news of a successful chain reaction was a virtual explosion to American scientists, and scientists lost no time in verifying the news brought by Dr. Niels Bohr. They found that it was true, and when Dr. Leo Szilard read Dr. Lise Meitner's report concerning Dr. Hahn's successful fission with uranium, he immediately perceived that a chain reaction with uranium could cause an atomic explosion. He knew the process that he had been working on in England was now practical and he was elated, but his elation was also accompanied by fear for the consequences were too gruesome to even think of if Hitler got possession of an atomic bomb.

Because of Dr. Szilard's heightened concern about the possibility of an atomic bomb in Hitler's arsenal, Dr. Szilard personally persuaded American physicists to keep mum and not to mention anything about fission to prevent the German's from learning any knowledge about American research concerning fission.

During the critical year of 1939, Princeton University was honored by having three brilliant Jewish Hungarian refugees (all did become naturalized) that had fled Germany early in the 1930s to get away from Hitler. While the three worked at Princeton University, (Dr. Leo Szilard, Dr. Edward Teller and Dr. Eugene Paul Wigner) their triune think tank was led by Dr. Leo Szilard and they planned to inform President Roosevelt that a super bomb could be built, and decided to use Einstein in their plan. But before entreating Einstein to lend his prestige in influencing President Roosevelt to consider an atomic bomb project, the method to achieve fission was explained to Einstein by Dr. Leo Szilard. Szilard carefully explained the fission process to Einstein and as Einstein became convinced, he said slowly, "I never thought of that."

Pictures were taken of this Szilard-Einstein meeting (shown on the History Channel); therefore, it is a documented fact of history and is important in order to give proper credit to the actual designers of the first atomic bomb and to reveal the scientists that did nothing in the development of the Atomic bomb.

It was Dr. Leo Szilard who originally conceived the method to cause a fission chain reaction, and it was also Dr. Leo Szilard that conceived the plan to use the fame of Einstein to make direct contact to the president in that summer of 1939. However, at that time America was not directly in a war and not directly threatened for in 1939 it was over two years before Pearl Harbor. Thus - Dr. Leo Szilard, Dr. Edward Teller, and Dr. E. P. Wigner entreated Einstein to become a part of a plan to notify President Roosevelt that a super bomb could be built.

According to Isaac Asimov (1920–1992) in his *Biographical Encyclopedia of Science and Technology* (and other sources) the records are revealed that not only did Dr. Leo Szilard explain the fission process to Einstein, but Leo Szilard actually wrote the letter to President Roosevelt dated August 2, 1939 that Einstein signed. Thus, that was the extent of Einstein's involvement in the Manhattan Project.

However, President Roosevelt could not initiate an atomic bomb project in 1939, for Congress would have had to be notified and would have had to appropriate the funds for the project. As America was coming out of the Depression in 1939 and America was not yet threatened by Germany or Japan, the federal government embarking on an atomic bomb project was next to impossible.

However, an event occurred in the course of Hitler's armies marching across Europe that was to later influence Roosevelt's resolve to have atomic bomb research started in America. Hitler's army seized 1,200 tons of uranium ore from Belgium, along with valuable atomic technology, and had it sent to Germany. In addition, all exports of Uranium from the Czechoslovakian uranium mines were halted. Also, in the late 1930s a chemist from Bakersfield, California, was unsuccessful at interesting members of Congress and President Roosevelt in a process that he had conceived to extract heavy water from ordinary water. This process is vital in the development of the atomic bomb, for a large quantity of neutrons is needed to enable uranium to be moderated when it is transformed to plutonium. Heavy water possesses the heavy isotope deuterium that contains a neutron in its nucleus.

However, instead of granting the chemist an audience in Washington, powerful (and unknown to me) people had the FBI investigate him, since he had sent a voluminous amount of letters to people in the government. Certain people in Washington believed the chemist to be slightly unbalanced mentally; but German agents had also heard about the chemist from Bakersfield, California, and went to see him. The result of that meeting was an offer by the German agents to the chemist to go to Germany to accept a position of being in total charge of a heavy water plant on the North Sea. The chemist accepted, and Germany was successful at forging another link in a plan to develop an atomic bomb.

The exact nature of the American chemist's contribution to Germany's plan to build an atomic bomb has not been discovered by researchers. Whether the American chemist's knowledge was a factor in Germany's war and occupation of Norway is also unknown by this researcher, but it is known that Germany commandeered the Norwegian heavy water plant and armed it with protective defenses. Also, whatever became of the American chemist remains an unsolved mystery of World War II.

However, American plans to build an atomic bomb gained new impetus after the bombing of Pearl Harbor on December 7, 1941. That devastating attack was a declaration of war on the United States, and America's limited involvement in helping Britain became an all-out involvement as Germany declared war on the America. Hitler had a military pact with Japan that tied Germany with Japan if Japan became at war with America. The December 7 Pearl Harbor raid was used to induce Hitler that Japan was at war with America. So at 3:30 p.m. (Berlin time) on December 11, 1941 the German charge d'affaires in Washington handed American Secretary of State Cordell Hull a copy of the declaration of war. Congress had no choice but to declare war on Germany and Japan.

## MANHATTAN PROJECT WAS BORN

Shortly afterward, in 1942, full war mobilization began, which included the project to build an atomic bomb, dubbed the

Manhattan Project. It was begun with an initial $500 million annual budget. The concept of an atomic bomb originated in the general Germanic European region and the Germans were at work with a program to build an atomic bomb.

As previously explained, neutrons are a necessity in the atomic process to obtain the fissionable element plutonium and to obtain neutrons; it requires vast amounts of electricity to separate heavy water from ordinary water. Since Norway had been invaded and its resources were used in Germany's war effort, the North Sea hydroelectric dam became strategic in Hitler's plan to build an atomic bomb. Norway was already extracting heavy water from ordinary water to sell on the commercial market. As soon as the Germans occupied Norway, they fortified the hydroelectric dam.

After the Germans fortified the heavy water plant along the North Sea, which had been named the Norsk Hydro by the Norwegians, they doubled the amount of heavy water extracted from ordinary water and began to store the extracted heavy water in barrels in its basement. A few Norwegian scientists escaped to England and informed British intelligence about the heavy water extraction plant, and it was placed high on the priority list to be destroyed. Two gliders were loaded with British commandoes and explosives and landed in Norway. One glider wrecked, killing all aboard, and the Germans caught the commandos of the second glider and executed them. In the third attempt, Norwegian commandos were trained in England and several attempts were made before one night raid was successful at placing explosives in the basement, where they exploded and destruction wreaked where the heavy water was stored.

However, within several months, the Germans had the plant operating again. American and British strategists were still determined to destroy the heavy water plant and heavy bombing raids by American and British bombers made the plant inoperable, but the bombing did not destroy the stored heavy water in the plant's basement. In response to this loss, senior German officials decided to have the heavy water shipped to Germany.

The first phase of that journey had to begin by shipping the container across the fjord by ferry boat. Norwegian intelligence people informed British intelligence and the Brits ordered the Norwegian commandos to destroy the ferry. It was a hard decision for the Norwegians, for innocent Norwegian people (and some of their relatives) would be on the ferry that was shipping the heavy water across the fjord. An explosive charge was placed on the lowest level of the ferry and it sank quickly after it exploded in the deepest water. Almost all aboard the ferry were killed; the heavy water containers that were not ruptured in the explosion sank to the bottom of the fjord and were expected to be there forever.

However, over 60 years later, divers went down to the sunken ferry and retrieved one barrel of heavy water. It was determined even if the total amount of heavy water in the ferry reached Germany, it was only one-tenth of the amount needed to make one atomic bomb. At that time, no one had that knowledge and the allies were closing on Germany. In retrospect, it was unnecessary to sink the ferry. A more detailed account of this WWII history can be found on Wikipedia.

However, when the ferry was sunk, no one knew how soon Germany scientists would have an atomic bomb. A few barrels of the sunken ferry were only partially full, and they floated to the surface. Germans retrieved these barrels, and it was presumed they had been sent to Germany. Whatever happened to these few barrels of heavy water is unknown; however a few weeks before Germany surrendered in 1945 American army personnel found an unfinished nuclear reactor in a bombproof bunker in southwest Germany. American scientists calculated that an additional 185 gallons of heavy water would have been needed to start production of plutonium that is necessary to build an atomic bomb.

**THE RACE TO BUILD AN ATOMIC BOMB GOES INTO HIGH GEAR.**
After Pearl Harbor, American leaders spared no cost in what became the race to build an atomic bomb before Germany. Los Alamos, New Mexico was chosen as the central place for research and the development of the atomic bomb. The military leader of the Manhattan Project was a West Point graduate and engineer, General Leslie Richard Groves. His closest military colleague that served as his deputy was General Thomas Farrell. The New York-born Jewish physicist Dr. Robert Oppenheimer (1904–1967), was appointed to the position of project manager at Los Alamos. A very close colleague of Dr. Oppenheimer's from Berkeley, Dr. Ernest O. Lawrence (1901–1958), also joined the Manhattan Project. Dr. Lawrence worked at Oak Ridge and became in charge of electromagnetic separation work that provided the fissionable material for the atomic bomb.

The five German-Jewish physicists that led the team at Los Alamos were: Dr. Robert Oppenheimer, Dr. Leo Szilard, Dr. Eugene Paul Wigner, Dr. Edward Teller, and Dr. Richard Feynman. Three additional Jewish physicists that helped with the development of the atomic bomb that had also fled to escape the Nazi regime were Dr. Niels Bohr, Dr. John Von Neumann, and Dr. Hans Bethe.

Other physicists involved in the project were: Dr. Arthur

Holly Compton (1892–1962), Dr. Enrico Fermi (1901–1954), Dr. Robert Serber (1909–1997), and many others including Dr. Louis Rosen, Joseph O. Hirschfelder, Joseph Hoffman, Louis Hempelman, Robert Bacher, Victor Weisskopf, Kenneth Bainbridge, Richard Dodson, Norman S. Ramsey, Leo James Rainwater, Robert Wilson, Norris Bradbury, Boyce D. McDaniel, the explosive expert Seth Neddermeyer, and his replacement George Kistiakowsky, plus several British scientists.

These were the scientists responsible for working directly on the project of building the first atomic bomb or were instrumental in the development of the atomic bomb. While the physicists Szilard, Teller, and Wigner were born of Jewish parents in Budapest, Oppenheimer and Feynman were born of Jewish parents in New York.

In the decade of the 1930s, the German physicists Otto Hahn and Fritz Straussman, plus the Jewish physicists Leo Szilard, Lise Meitner, Niels Bohr, and Enrico Fermi from Italy, were among the European developers of the fission process. However, in America, as the decade of the 1940s began, at the University of Chicago, nuclear physicists led by Enrico Fermi were working on causing a controlled atomic reaction. The concept of causing an atomic explosion by fission was proven by Enrico Fermi and his scientific team on December 2, 1942. About forty-two scientists were involved in making the first atomic pile under the stands at Stagg Field in Chicago. Among the scientists in addition to Enrico Fermi were H. L. Anderson, Arthur H. Compton, Leo Szilard, Albert Wattenberg, George L. Weil, E. P. Wigner, and W. H. Zinn.

However, the atomic pile experiment conducted in Chicago was designed to control the fission process. It is the same principle used at nuclear power generation plants. If the fission process is allowed to continue without being controlled, an atomic explosion would occur. It's easy to write these words; however, in order to make an atomic explosion, huge facilities were required to mine the uranium, process the uranium and build a device that would explode.

The first test explosion was conducted at a site called Trinity, in the desert near Alamogordo, New Mexico, on July 16, 1945. The explosive force generated was equivalent to 20,000 tons of TNT. The total weight of the bomb was about 10,000 pounds, yet the center pit of fissionable material weighed about two and a half pounds and was only about the size of a baseball.

Scientists stood afar behind cement and sandbag protection, and as the bomb exploded, it vaporized the metal tower upon which it was mounted. The scientists wore eye protection; yet the tremendous brilliant ball of white light transfixed them for several moments. As they recovered from their astonishment, Dr. Robert Oppenheimer made one remark, saying, "It worked."

However, the American military questioned whether the atomic bomb would ever be used, for after the greater portion of Tokyo was destroyed by conventional bombing, chiefly with incendiary bombs, American strategists believed that Japan would sue for peace. But the Japanese war-crazed warlords were so deeply afflicted by a war craze that the danger of Japanese obliteration was chosen rather than surrender.

Then twenty-one days after the first test atomic explosion in New Mexico, on August 6, 1945, the first atomic bomb (a plutonium gun type device) code named "Little Boy" (a changed name from "Thin Man" after President Roosevelt) was dropped on Hiroshima from a refitted B-29 Super Fortress piloted by Colonel Paul W. Tibbetts Jr. The plane was named the *Enola Gay*, after Colonel Tibbets' mother.

Japanese leaders were warned to surrender, but they continued to refuse or heed the American demand for unconditional surrender because the warlords craved one last battle with the American forces that was planned for the invasion of Japan. The invasion at the Normandy beachhead cost 10,000 American lives and an invasion of Japan was estimated to cause the death of nearly a million Americans and nearly obliterate Japan. Therefore, the option to invade Japan that would cause such a loss of human life and destruction was not favored over using atomic bombs. So the first atomic bomb was dropped on August 9, 1945, on Hiroshima. The Japanese were again warned to surrender and did not respond to the ultimatum. Thus, the second atomic bomb (a uranium-235 implosion type) code named "Fatman" (after British Prime Minister Winston Churchill) was dropped on August 12, from another refitted Super Fortress named *Bock's Car.*

After the second atomic bomb was dropped, foolishly, a third atomic bomb was planned to be dropped on Tokyo. This was an absolutely crazy idea, however, for the benefit of Japan and humanity before it was dropped the Japanese emperor realized the devastation that would come if Japan would continue the war. Consequently, he refused to listen to any more war plans of the warlords and arranged for a Japanese surrender, which came on August 15, 1945.

If Tokyo had been bombed with an atomic bomb, the leading people necessary to surrender would have been killed. This would have resulted in more atomic bombs destroying Japanese cities. Fortunately for Japan and the world, the emperor was alive and made the wise choice to surrender.

Using the atomic bomb saved hundreds of thousands of lives and saved Japan from an onslaught of around-the-clock conventional bombing, more atomic bombing, and near obliteration.

After the war, Dr. Lise Meitner became a Swedish citizen, and in 1958 she moved from Sweden to Cambridge, England. In 1966, the American Atomic Energy Commission awarded her, Dr. Otto Hahn, and Dr. Fritz Straussman a Fermi Prize for that year.

For Dr. Lise Meitner, it was an impressive award, for not only did she have a vital role in bringing atomic energy to use, but she was the first woman to be awarded with such great acclaim in the history of atomic energy. She was fully devoted to science and never married, departing from this Earth just before her 90th birthday.

Therefore science scholars, concerning the key scientists that devised and developed the atomic bomb, the history that is given here reveals the identity of the key scientists and their genealogy which was Jewish and German. The Jewish scientists were of the Judaic faith but through inter-marriages had a high percentage of German genes.

The source data that I drew from came from numerous historical references including, Isaac Asimov's *Biographical Encyclopedia of Science & Technology*, Funk & Wagnalls, *Americana*, and the World Book Encyclopedias, plus the CD ROM from Corbis entitled, *Critical Mass*, a TV documentary from the History Channel, and the websites tacair-press.com, atomicarchive.com, and the Smithsonian magazine.

If you ever believed, in error, that Einstein helped to work on the atomic bomb, now you know the truth. Einstein did not help to develop the atomic bomb. That bit of history was important, for it clarified the point that Albert Einstein had nothing to do with developing the atomic bomb except to sign Leo Szilard's letter that was sent to President Roosevelt. However, it took us far and away from my specific focus of present-day opinions that are regarded as laws. Let us continue with an in-depth analysis and an exposé of false ideas. End of the history about the development of the atomic bomb.

### GENERAL RELATIVITY

Why general relativity has not been dissected in the last almost 100 years is somewhat of a mystery to me. Possibly it is due to a weakness in in-depth analysis of some scientists and reluctance of other scientists to dissect Einstein's general relativity. However, since it has not been done, it stimulated my curiosity to totally dissect general relativity.

After I understood that some of Einstein's foundation postulates were not facts but were his personal opinions and were false, I was amazed and still am to this day as to how the scientific world could have accepted such a gobbledygook theory. Truly, several of the postulates that make the framework of general relativity have no basis in fact. For example, the postulate that states when matter is accelerated to near the velocity of light it will shrink in the direction of its movement, *gain mass* reaching a limit at the near or 99th percent of the velocity of light.

Einstein did not give a rational to explain *why and how* this could occur, except to accept the erroneous theoretical conjecture of Lorentz.

*Gain Mass*—mass is energy and energy—so far cannot be created, so where was this new extra energy supposed to have originated? Einstein had no rationale to explain this supposed "acquiring energy" idea. Einstein was content to rely on an unproven conjecture of Lorentz. Einstein simply believed this conjecture on the grounds that Lorentz said so, and Lorentz is a smart man, so it must be true.

## PLEASE CAREFULLY CONSIDER THIS GAIN MASS POSTULATE.

Not only was Einstein deluded, ever since general relativity was accepted to be true by organized science, every supposed-to-be smart physicist that believed in Einstein's fallacious idea was also deluded. His fallacious theory did not delude the Nobel Prize awarding committee.

A ninth-grade science student should be able to perceive the flaw in the "gain mass shrink in the direction of movement" idea. Consider the following: mass is energy, not all energy is mass but all mass is energy. At the time of general relativity's creation and up to the present until researchers will finally understand that "energy dissolves" the idea that "energy cannot be created or destroyed" was and is a foundation principle of modern science.

Therefore, since energy cannot be created and a body of mass when rapidly accelerated to near the velocity of light is supposed to gain mass, *or to gain energy,* where in dark free space could mass absorb energy that does not exist and cannot be created. Make a deep note of this in your mind, for we will return to this in depth perspective.

This "gain mass shrink in the direction of movement" idea is for science fiction or funny books for energy—whether it be in the form of forces or mass, cannot be created from nothing in the physical universe.

From my unbiased research, I discovered from each and every search into aspects of general relativity that *Einstein had ignored obvious facts* during his creation of a generalized theoretical framework *in favor of theoretical propositions presented by other revered men of science.* Einstein drew from the works of other theoreticians in a manner to encompass light, the nature of energy, subatomic particles, solar system dynamics, and gravity into one grand theory. He deluded himself by believing that his theoretic material that was taken from other theoreticians could not be wrong. By this great mistake, he misled all that believed in his laced-together false concepts.

As I dissected general relativity, I became amazed that some researcher had not made the same discoveries that I had made. I have read some works of theoreticians in California that do not believe in the claims of Einstein, as did some of his own peers, nor the Nobel Prize review committee. Yet, I never read a point-for-point, counterpoint exposé of the postulates of general relativity, and I never read where any scientist perceived a rotary gravity plane with an extended gyroscopic plane as the force that holds and drives the planets around the Sun.

After I dissected general relativity and understood its flaws, I also speculated at the possibility that a type of enchantment had gripped the scientific world that prevented leading scientists from objective analysis. This possibility seems plausible for the first several decades after the initial introduction of Einstein's General Relativity; however, as decades passed I believe the leading scientists of the world became doubtful of Einstein's theory and were ashamed to allow the world to know that they have been duped and were teaching false theory for decades. This icon of science, Albert Einstein, whose name is revered by millions if not billions of people has had statues built of him at prestigious places by ardent believers of his false concepts and erroneous postulates. Universities, colleges, technical schools, and public schools all teach his false concepts as profound truth.

Yet, Einstein's work has in no way changed the lifestyles of humanity or in any way altered our environment, nor has it enabled us to, in any manner, capitalize upon our environment, while the three greatest people next to Christ, whose work changed the world are relegated to a second or third place in the importance of contributions. These men are: Johannes Kepler, Michael Faraday and Thomas A. Edison. Thomas Edison himself and the team that he hired brought forth astounding works that changed the life of humanity and enabled us to capitalize upon our environment in manners unparalleled in the history of humanity.

For a consideration of comparison of the outstanding contributions of Thomas Edison v. the contributions of Albert Einstein, thoughtfully consider the following beliefs and claims that illustrate Einstein's psychological profile and how some of the ideas developed by others that he erroneously accepted were incorporated into the theory of general relativity.

By Einstein's own admission, his perception and mechanical skill or innovativeness was minimal. Due to this, he never researched the following:

1. Why do the planets revolve around the Sun in precision movement according to their distance from the Sun?

2. What is the true shape of each planet's orbit?

3. Why do all the planets revolve in the direction of the Sun's rotation?

4. Why do all satellites revolve in the direction as their controlling planet's rotation excepting Neptune's Triton?

5. Why do the radial arms of a spinning galaxy revolve in the

direction of the rotation movement of the galaxy's central Sun?

6. What causes the suns of a Globular Cluster galaxy to be randomly located around the central Sun? In addition, Einstein was unable to comprehend a universe created by a Supreme Mind. Atheism was his philosophy.

7. Einstein believed inertia causes planets to revolve around the Sun.

8. Einstein could not understand why Mercury, Venus and Earth were not pulled into the Sun billions of years ago. To account for this Einstein concocted an idea that gravity is a condition of space where space is non linear 'warped and deformed' in the vicinity of matter.

9. His belief stated that planets simply move by inertia and take the shortest possible path in a non-linear "warped and deformed" space. From this false idea he incorporated his idea of troughs. Although, he never submitted a rational to explain *why* space would be "warped and deformed" and *why* troughs would be formed.

10. From the fact that man's measurement of motion binds motion and time together, Einstein developed a new term "Space Time Continuum" for measuring intervals of motion that pertains to galactic continuous motions.

Measuring how long or how much time is required for certain movement in atomic and molecular motions takes for granted that research has bound time and motion together. Although, this being true for atomic or inorganic substances, it cannot be true for most organic substances. Simple organic life can be frozen for millions of years; this condition causes time to be arrested with its arrested motion. But extreme cold or heat will kill complex organic life forms that have developed outside of extreme conditions, for the programming of metabolic functions has been deeply set and is susceptible to shock.

11. Einstein correctly stated that time is bound together with all movement in the universe.

His emphasis on time has confounded all superficial researchers since general relativity was introduced to the world. The fact that atomic time slows for water when it freezes and quickens when water is heated enough to cause water to become a vapor is no earth-changing revelation. This fact was taken for granted before Einstein focused upon the time aspect of motion.

Ancient Chinese, Chaldeans, Mayans, and fastidious astronomers such as Tycho Brahe knew the importance of the date and time of day when recording the position of a heavenly body. For centuries before Einstein, and to this present day, researchers note the position of heavenly bodies and the time when their position was recorded. No one ever suggested the gobbledygook idea that their movement occurred within a collectively called "Space-Time Continuum."

Although it is true that all the astronomers before Einstein, except Kepler, did not need a weird theory to associate planet's stable orbits and why they moved in the solar system; however, Einstein's idea of inertia being the force responsible for planetary movements and his idea of a "warped and deformed space in the vicinity of matter" that created troughs to maintain their orbital position (of course without proof or an rational) became a fascinating idea.

After Einstein's special theory was proven to be correct, i.e. light bends as it passes a star, he gained respect and stature in the scientific community. The result of a chauvinistic mind set to those craving to know truths beyond the frontier of knowledge, they were suddenly given a new theory from the man that proved "light bends as it passes a star."

The combination of incorporating time as an additional coordinate, plus fancy names in *General Relativity* deceived writers of the London *Times* in 1919, and from their ignorance they deceived millions of people in their zeal to sell their daily paper.

12. Perhaps the electron does have an inner action, Einstein wondered and mused aloud.

13. Einstein firmly believed that energy cannot be created or destroyed only transformed.

14. Furthermore he stated: the ether does not exist and it would be better if we never hear its name mentioned again.

15. Magnetism also was another force that was supposed to simply be a condition of space where space is "warped and deformed" in the vicinity of matter.

Fortunately, the researchers that developed super magnets and super-super-magnets proceeded with their research with their basic concepts rooted in the molecular theory of magnetism.

Super magnets were brought into existence by alloying aluminum, nickel and cobalt while super-super-magnets proved to be more difficult, but the Swiss were helpful in this research, and it was found that samarium atoms had a football shape and could fix an alloy into a permanent polarity arrangement when properly 'heat treated'. Although the most powerful magnets were created using neodymium in an alloy with iron and boron.

However, without theorizing about the unknown, it is inconceivable to me how known phenomena that is given here was completely ignored when Einstein or any theoretician attempted to weave together diverse concepts into one theory. This is not only especially true since the counterclockwise rotation motion of the Sun and the counterclockwise revolution motion of the planets and their precision movement according to their distance from the Sun was a well known fact, In addition, the general operating principle of an AC transformer was known.

The nature of electromagnetism has baffled many researchers and prompted Arthur Holly Compton to say, "If waves be particles, why might not particles be waves?" By posing that question, Dr. Compton actually revealed the nature of particles, albeit with an interrogative posture.

To all scholars reading this book: to understand how this principle operates, read the "Particle Wave" at the close of these manuscripts. This was originally copyrighted by myself in 1963. Following are the irrefutable definitions of terms that will redefine physics of the nineteenth and twentieth centuries and establish a new physics and a new astronomy for the twenty-first century.

**TERMS DEFINED**

Energy *is the capacity to perform work.* It is a quality that is derived from harnessing physical conditions. The four classes are:

Inertial Energy—can be either *static* or *dynamic* and pertains to stationary or moving mass. These are further defined as a body at rest or a body in motion. Thank you, Isaac Newton—first law. The *static* and *dynamic* nature of inertial energy is further defined as either *potential* or *kinetic*, i.e. a rock stationary on the ground, a rock moving through the air, a coiled spring, a spring unwinding.

Atomic Energy—or the capacity to perform work is accomplished by atomic fission or fusion, i.e. an atomic bomb, thermonuclear reactions on a star.

Energy Inherent in Non-Physical Forces—these include electromagnetic forces, magnetism, gravity, and heat.

Energy Inherent in Physical Forces—this includes ocean currents and windmills, or a revolving cloud that generates or produces lightning, and maybe, I stated maybe by what is erroneously referred to as Cold Fusion or the latest rave, Bubble Fusion.

Work is overcoming resistance by causing or producing a change to matters state or condition or a change to magnetism and all electromagnetic forces.

*Energy Can Be Stored, Transformed, Dissipated, or Dissolved*
Energy can be stored, transformed, dissipated, or dissolved. Dissipated energy is energy that is released by a nonharnessed device. Dissolved energy occurs when electromagnetic forces are not harnessed to perform work. Electromagnetic waves and the minute energy concentrations within the wave structure may exist for billions of years without being dissolved. However, energy in this state cannot exist forever and when the wave structures is eventually broken, the energy inherent in stable

interactions within the wave structure will cease and will be absorbed by the primal etheric source that gave it its original identity.

## Ethons

Ethons are the basic primal substance of electromagnetic forces. Ethons are actually particles of a pervasive primordial substance in the universe called the ether. These primordial particles of the ether do not have a charge, that is, they are neutral in charge. They only acquire a charge when they begin to swirl. Also there are different grades of ether and I have named the building blocks of electromagnetic radiation or ether particles—ethons.

## Light Energy Dissolves

As the 'light energy' radiates through the ether and it could be for billions of years, the light waves will eventually encounter obstacles —that is, the light contacts matter. If the light energy is not absorbed by matter, the contact to matter causes the 'light wave structure' to be interfered with to the extent where the 'wave structure' is broken. When this occurs the light ceases to exist and the etheric particles or ethons of the light wave are absorbed by the ether.

## The Origin of Electrons

Electrons are formed from electromagnetic energy by various devices where a condensing of phases of electromagnetic energy by atomic swirling fields (eddy current or vortex) creates or forms stable energy particles or electrons. Two graphic illustrations are given here using light and electromagnetic force to convey the structure of electrons.

Electrons can be found naturally in cosmic showers and

in radioactive decay. In addition, heat, pressure on crystals (thermal couples, and piezo electricity) cause a flow of electrons if a way is provided for their transmission. However, the greatest sources of electrons are generators, alternators, solar cells, and chemical actions.

*The Nature of Electrons*

My analysis of the nature of the electron has revealed the sub-atomic particle to be composed of a phase of electromagnetic force. Solar cells convert light into electrons, generators and alternators form electrons from cutting across magnetic field. I repeat that worn cliché again: a picture is worth a thousand words, and in compliance with this truism, graphics are presented that illustrate the inner nature of electrons.

Electrons are continually radiating "fine energy particles" that are constantly being replaced in their inner workings by their own etheric vortex. This concept revives an ancient concept and a concept held by latter renaissance science or the early modern science known as the pycnosis concept from the Greek word *pycnosis*. It was chosen by theoretical physicists to convey "a vortex in a thick or dense medium."

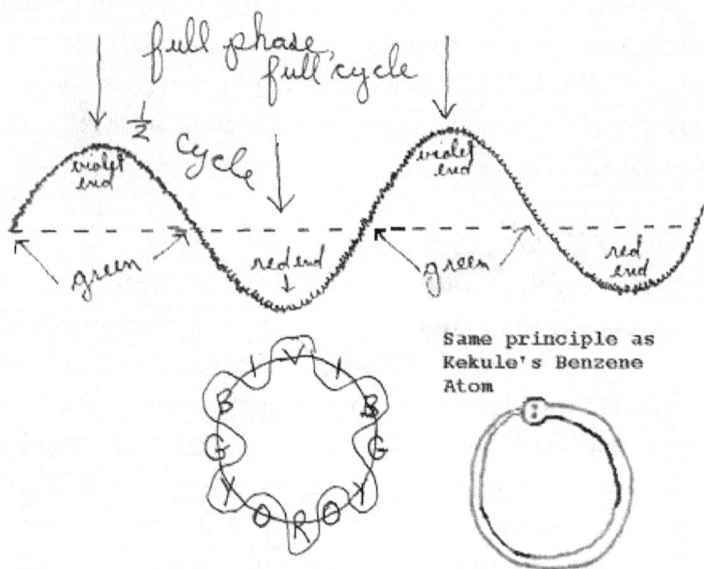

full phase,
full cycle
½ cycle
violet end
green
red end
violet end
green
red end

Same principle as
Kekule's Benzene
Atom

Rene Descartes (1596–1650) favored this concept as did, Lord Kelvin i.e. William Thomson (1824- 1907). Lord Kelvin was the last renowned scientist, until myself, to accept and push the belief of the ether. A history of Lord Kelvin can be read by going to Google and researching Irish scientists. A great innovator, but as a theoretician you the reader decide. In 1900, he stated to the British Association for the Advancement of Science, "There is nothing new to be discovered in physics now. All that remains is more and more precise measurements."

In addition, he stated, "Heavier that air machines are impossible"—Radio has no future"—"X-rays are a hoax." I have not discovered the reasoning for his acceptance of the ether. Perhaps the ponderous truth of a primordial substance came through his theoretical excursions.

*Free Electrons and Electrons in an Electrical Current*
The electron's nature is no longer elusive, nor is its shape. An electron in free space resembles a semi-spherical spiral galaxy. Since electromagnetic waves in free space propagate at the well-known rapid rate, it can easily be understood. When a completed cycle from the least energetic as red in the light wave to the most energetic as violet in the light wave is formed into a circle or semi-spherical shape the created or formed electron is very motive.

Electrons in an electrical circuit can be either formed or dissolved in an AC step-down or step-up transformer. In a step-down transformer, EMF, or voltage, is reduced and new energy particles or electrons are brought into being or amperage is increased. Or to paraphrase, in the building and collapsing field in a step-down transformer. Electrons are formed from EMF—thus the increased number of electrons is increased amperage while the decreased EMF is decreased voltage.

Conversely, in a step-up transformer, electrons give up their identity, that is great numbers of electrons cease to exist - are dissolved—from an electrical circuit that were measured in amps and coulombs. The dissolved electrons are formed into EMF or voltage during the building and collapsing field between the primary and secondary coils.

Coulombs. The above graphic and the following graphic assists in understanding the structure of electrons, and how electrons are formed and dissolved in a transformer and are formed into EMF or voltage during the building and collapsing field between the primary and secondary coils.

Please be sure that you understand what happens inside a transformer from the perspective of how electrons are formed and dissolved. If it is not clear, please carefully study the transformer graphic and read the text and think about it if at first you do not understand. Although for further clarification, a detailed explanation is given in the following paragraphs that will carefully explain that electrons are composed of a phase of the electromagnetic wave.

However, before the electron is dissected in the next few paragraphs, it is important to clarify the difference between direct and alternating current.

Texts that explain the nature of electricity explain how direct current flows continuously in one direction as water flows through a hose and alternating current does not flow continuously but shuts off and turns on according to the poles in the (alternator) generator that governs the number of cycles. Therefore alternating current flows through wires as though you hold a pistol grip on a water hose and squeeze it off and on quickly. The water will shoot out in intermittent surges and that is exactly the way that alternating current flows. Alternating current flow in surges; nothing alternates in an alternating current circuit unless two separate circuits are created to carry two separate surges.

Electricity does jump from a more highly charged place or where electrons or electromotive force is more concentrated to a less concentrated place. This type of electricity has been named static electricity. But electricity produced by a generator or a alternator is dynamic electricity and it flows in transmission lines that enable the current to flow.

Electrons and electromotive force are both a dynamic force therefore a push force is associated with them. Direct current that originates from solar cells, or a battery leaves the negative terminal then follows the transmission medium provided in a continuous flow as a long train to the point of escape or point of transformation or the point of resistance and returns to the positive terminal. Yes—it returns to the terminal minus the electrical energy that was lost during transmission from line resistance.

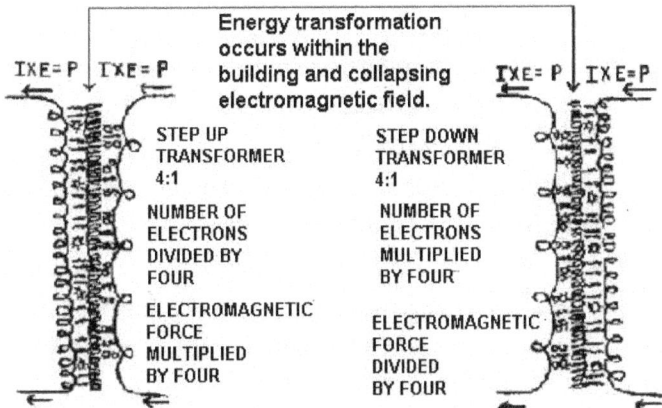

Energy transformation occurs within the building and collapsing electromagnetic field.

$IXE = P$   $IXE = P$                    $IXE = P$   $IXE = P$

STEP UP TRANSFORMER 4:1

STEP DOWN TRANSFORMER 4:1

NUMBER OF ELECTRONS DIVIDED BY FOUR

NUMBER OF ELECTRONS MULTIPLIED BY FOUR

ELECTROMAGNETIC FORCE MULTIPLIED BY FOUR

ELECTROMAGNETIC FORCE DIVIDED BY FOUR

CERTAINLY ELECTRONS HAVE AN INNER ACTION

Electrons are dissolved and reformed where the building and collapsing field motion reformulates electrons and electromagnetic force according to the number of windings in the secondary coil.

Electrons are very motive spherical shaped balls of electromagnetic energy, their shape can adjust according to forces acting upon them.

Originally copyrighted in 1963 and 1984

Jack R. Truett Sr.
Research Analyst

While in a DC generator the current is produced to flow in one direction then reverse and flow in the opposite direction, however a commutator in a DC generator enables the current to continue to flow in the same direction. An AC alternator does not have a commutator and generates or creates two distinctly separate currents, thus creating a current that alternates in surges between two transmission lines. Thus the current flows in alternating motion from one wire to the other, however, not in a continuous flow, but in surges. One surge starts, reaches its maximum, and diminishes as it ceases to flow and the field associated with the current collapses. However as that surge ceases another surge commences, reaches its maximum, diminishes and its field collapses. Thus instead of being a continuous or direct flow of electrical current there is an alternation from one surge to another and hence—alternating current IS really continuous surges of electrical energy.

Each distinctly separate surging current that flows from an AC alternator has a negative and a positive terminal, thus instead of having the wire connected to the negative and the positive as a generator, one wire is used for the return for both circuits and therefore only three wires are used because the return serves as a return for both circuits since only one circuit at a time is powered by surges of electrical energy. These two separately created circuits are commonly referred to as two legs.

In your home, each receptacle has a live (electrically energized) or hot wire (the black wire) that carries the pushing electrical current and a neutral wire (the white wire) that pulls the electricity. Therefore this current does flow in a similar manner as direct current from the perspective that electricity flows in one direction. Although whether it be an AC or DC current, a push and pull condition must be established.

The 110-volt current in your house is coming from one leg of the three-wire circuit, therefore the transmitted electrical current into your house is only half of the total two leg circuitry and thereby that one circuit carries current in one direction, although rapidly in surges.

Consider the following: In a single-phase, 60-cycle current, a rotor revolves 60 times a second and passes one field. Thus there are 60 surges of electrical energy flowing every second. Therefore the flow of electricity is actually shut off and on 60 times a second. Thus in the given example an incandescent light bulb would be actually turning on and off 60 times a second, but it would be so rapid the dimming from highest intensity to off then rising in brightness to maximum is imperceptible.

*But in a transformer, the turning on and off causes a corresponding building and collapsing of the electrical flow's field.* As the total energy in the electrical flow, that is, the electrons and the electromotive force collapses, a building again begins but depending upon the windings in the transformer, the building force will become electrons and electromotive force in direct relation to the number of windings between the primary in secondary coils. However if an alternator is built that has more than three fields (four, five or up to twelve) for the rotor to pass in one revolution, the current produced has been named polyphase. This type of current is used for transmitting high voltages, although transformers cannot be powered by a polyphase current because transformers cannot transform electrons to electromotive force or electromotive force to electrons with a polyphase current, because the building and collapsing field is too rapid to cause the transforming effect. Polyphase current resembles direct current.

*In a step-up transformer, some electrons give up their identity to become electromagnetic force. In a step-down transformer, some of the electromotive force gives up its identity to become electrons.*

Please read the above two sentences again. Do not proceed until you fully understand this principle. The coiled wires in the primary and secondary coils of a transformer are intertwined, yet without touching since the insulation prevents this. So, the area where the coils occupy space has field or an electromotive force field within it and surrounding it and it is this field that enables the collapsed surge of electrical energy (EMF and AMPS) or voltage and electrons to reformulate into a new generation (a new surge) of electromagnetic energy (voltage) and (amps).

In the future, when a slow-motion video is created for students, it will enable the entire class room to view this collapsing and building field and reformulation of EMF and electrons in a transformer. All will be able to see electrons becoming dissolved in the ether field that pervades all space within the space between the primary and secondary coils as the surge of electrical current ceases and therefore collapses.

Then as the building commences, all will be able to see electrons begin to form into spherical shapes, i.e. spheres or balls of energy, then move to the secondary coil in equal quantas or amounts according to each winding of the secondary coil.

Simply by changing the number of windings in the primary and secondary coils, an AC current can be stepped up or down. That is: the number of electrons and the electromotive force can be inversely changed. In addition, special transformers can be built that has several separate coils comprising the secondary coil. By this arrangement, several individual circuits can be created to carry differing amounts electric power - power or P is equal to E x I. The letter *I* represents the intensity or amperage (electrons) of an electrical current. The intricate nature of electricity will be further revealed as you read.

Electrons have an extremely small mass. The mass of an electron has been calculated to be about one two-thousandths of the mass of the hydrogen atom and from another perspective the mass of a proton is calculated to be 1,840 times the mass of the electron. Although atomic electrons differ slightly in mass, the densest being in the first electron shell and the least dense being in the seventh electron shell.

Electrons are spherical in shape because all electronic particles, Suns and stellar bodies that were created from Sun material are spherical because this is the most efficient shape for the greatest amount of mass to be held in the least amount of space. Although, electrons being more tenuous than particles of greater mass, they will more easily adjust shape according to the forces acting upon the electron/s.

The building and collapsing electrical flow in a transformer that dissolves and creates electrons is so simple to comprehend. But Werner Karl Heisenberg, Louis De Broglie, Erwin Schrodinger, Paul Dirac, and others overlooked the operation of building and collapsing field in a transformer. When they tried to understand the nature of the electron, they chose to equate spectral patterns with the nature of electrons. This association came from the experiments of bombarding electrons on thin metallic films, and with this experiment electrons disappeared or dissolved and interference patterns appeared.

This easily understood concept of a building and collapsing field in a transformer that actually dissolves and creates electron particles also eluded Albert Einstein. Because of this, any of Einstein's theoretical expansions related to the nature of the electron were destined to be formulated on an erroneous platform or basis.

Einstein is not the only scientist that developed false or erroneous theories in our long history and he may not be the last, but please try to appreciate my position. If I simply present my findings without a rationale and without a history of how small erroneous concepts were added to others until great erroneous concepts were accepted as truth, scores of questions will immediately come to the fore and will have to be answered. Thus, this chosen format answers the questions that would naturally accompany a new science.

I have tried to be very thorough in my searches in order to explain, how and why present beliefs came to be accepted as the gospel truth.

Due to the propagating nature of electromagnetic energy, electrons are self-motive energy particles. Electrons' motivity is generated by an inner action or inner working of their electromagnetic composition, hence electrons are motive.

The motive nature of electrons causes electrons to resist being held in an enclosure. When a condition is created that restrains electron's motive nature, an electrical pressure is created. All electrical devices operate by motive electrons and the electrical pressure they create.

When electrons move in pairs through space or through matter, their meshed fields causes a clockwise movement of the one electron while the other moves in a counterclockwise direction. The same principle operates in the movement of two spur gears, they move in opposite direction when their meshed gears turn.

Electrons and EMF in an electrical current are both transformed into another form of energy when electricity encounters a resistance. For example: when electric current passes through devices devised to create lighting, electrical energy is transformed into photo electromagnetic energy or simply light energy. The total wattage drop in an electrical current is the exact amount of energy that was transformed into another energy form, eventually dissolved and absorbed into the ether.

When water is drawn from any spigot, it is used for a purpose and cannot be put back into the pipes for the water company to resend it to the same or another customer. Similarly, when electric current is used, for example lighting, it cannot be gathered and put back into the wires for the electric company to reuse it.

The kilowatt-hour meter at each house measures the amount or quanta of electromotive force and electrons (together called wattage) that is used each hour by your contrivances. The wattage (power) used is the electrical energy measured per hour (watt hours) that has been transformed onto heat, light, and overcoming resistance to operate electric motors. In each case, the *electrical energy* (measured in watt hours) that is bought from the power company is actually *taken from the circuitry*, transformed into another form and then in sequence, either is dissipated or is dissolved into the ether.

If you shine a flashlight into a room painted a reddish color, some of the light energy in the red end of the light will be absorbed by the walls, transformed to heat and eventually dissipated. The light energy in the blue end of the light energy will be dissolved into the ether. I am confident that all my readers are comfortable with the concept of dissolving energy as illustrated in my dissolving energy box.

**ETHONS ARE INDESTRUCTIBLE AND ARE THEREFORE CONSERVED**

As given before, ethons of the ether are conserved and could be absorbed into etheric cosmic streams eventually helping to feed a "black hole" if the ethons form a vortex.

Black Hole theory is still evolving from the Nobel Prize-winning theoretical work of Subrahmanyan **Chandrasekhar** (1910–1995). He was a physicist from India that had received his PhD from Cambridge. In 1936, he moved to America, where he joined the faculty at the University of Chicago; later, he also worked at Yerkes Observatory. During his life, he wrote on many scientific subjects.

Although a black hole concept was first suggested in 1795 by the French mathematician Pierre-Simon, marquis de Laplace (1749—1827), some erroneously believe that Einstein created this concept when in truth this is another aspect of Einstein's theoretical conjectures that was false. Einstein believed the idea of a collapsed star becoming so dense that light could not escape and it became a closed sinkhole was a preposterous idea.

In 1967, twelve years after Einstein died, the famous American physicist, John Wheeler, used the term *black hole* in a speech at Columbia University. Since then the interest in black holes has intensified.

### BLACK HOLES EXPLAINED

When a star eventually dies, or when it uses all of its fusion material (hydrogen), the densest chemically inactive particles (neutrons) that were attracted to the center of the star's core, remains the chief particles of the star. Thus, a neutron star is a thermonuclear dead star. Some neutron stars are known as pulsars, because some light is emitted in pulses, which can be caused by at least two conditions.

One cause of pulses of light is caused by hydrogen atoms being attracted to the neutron star and immediately being transformed by fusion into helium atoms. The second type of pulses of light comes in regular intervals. The most respected theory about these phenomena is the possibility of an interaction of its magnetic and electromagnetic forces due to the neutron star's rapid rotation or spinning motion. However, if the gravity of a neutron star is so powerful that the escape velocity is greater than the speed of light, the star cannot emit any light and the star temporarily becomes a black hole.

........................................................................

# I submit in this work that neutron stars do have a life cycle.

........................................................................

The period of time for a neutron star's life cycle may be millions or billions of years until it goes through its phases and eventually feeds the creation of new matter when its crushing gravity force reaches a critical point and new matter is spewed forth. Neutrons are stable within an atom and are stable within a black hole, but when neutrons are free in space, they quickly transform into a proton and an electron. Therefore, every neutron in a black hole has the potential of becoming a hydrogen atom.

This concept is submitted by this Research Analyst, but I'm sure each of my readers that are scientifically inclined is aware of the phrase "the only thing constant is change." This wise old saw has been paraphrased by modern writers but its origin is dated to about 700 BC. Therefore, from our observation of nature in constant change, it would be inconsistent if neutron stars would remain fixed forever in their nuclear condition.

It is quite reasonable to speculate that the rapid rotation or spinning of a neutron star creates a gravity vortex that has a tornado affect, by pulling in all particles within its surrounding space. In 1918, two Austrian physicists, Josef Lense and Hans Thirring, suspected that neutron stars could create a gravity vortex, but were so enthralled and deeply deceived by Einstein's theory of general relativity that relativity vernacular was used to try to explain the gravity vortex. Later the term *frame-dragging"* was created to explain the tornado affect of pulling in all particles within the surrounding space of a neutron star.

The super-super gravity field of a spinning pulsar is an abbreviation for a neutron pulsating star.

A spinning neutron star certainly does create a powerful Aust. In this case, the powerful pulling in gravity force with its magnetic force pulls hydrogen atoms to its surface where through Thermonuclear Fusion they are transformed into helium atoms and pulses of light is seen.

In addition, its highly possible that ether particles in magnetism is pulled into its south pole region where the magnetic force with electromagnetic forces are crushed into protons by the tremendous gravitational force and this could cause additional pulses of light to be emitted.

In addition, the name of another neutron star that emits radio waves was given an acronym name, quasar, for quasi-stellar radio sources.

Why pulsars do not emit the radio waves as quasars do has not yet been fathomed. It is logical to speculate that radio waves come from a transformation of energy within a neutron star. My own speculation as to the life cycle of a black hole that has been given here differs from secular science just as I differ from organized science in my ten claims given in the earlier pages of this volume.

The term *planet* is defined by the Kant-Laplace nebular hypothesis theory, and Kepler's laws. Immanuel Kant,

(1724–1804) and Pierre-Simon Laplace, (1749–1827) both speculated that our Sun was formed by swirling gases. I believe the Kant-Laplace nebular hypothesis theory is a false theory. It is possible for an open space area of the cosmos to have had swirling gases slowly form a spheroid shape until gravity attraction caused fusion to ignite the ball of gasses and a sun is born. However, galaxies are composed of millions of suns or stars and to attribute the birth of the galaxy's suns to a collection of swirling gases is not only illogical but preposterous. Numerous great scientists have submitted different viewpoints to account for the creation of our sun and the planets. For those that are interested in the history of theories concerning the creation or our sun and the planets, go to a scientific encyclopedia or Google. Search for the works of Emanuel Swedenborg, Immanuel Kant, and Pierre-Simon Laplace.

In 1734, Emanuel Swedenborg was the first to submit a theory concerning the birth of our Sun and the planets of our solar system. In 1775, Immanuel Kant submitted another theory that was built upon the idea of swirling gasses. Following his theoretical work, in 1796 Pierre-Simon Laplace submitted a slightly modified version of Kant's theory. The combined work of Kant and Laplace has been accepted for over two hundred years and is known as the Kant-Laplace nebular hypothesis theory.

We are now in the 21st century, and I don't believe the focus of any scientist is on the creation of our galaxy's suns. It is a generally accepted fact that our galaxy was created first. From this perspective, I will submit my understanding about the creation of our galaxy, its suns or stars, and planetary systems.

When a galaxy is created, physical matter is created and physical matter is energy. Therefore, a process occurred that created energy on a massive scale. While researching into deep space, the Hubble Telescope has viewed a gigantic explosion and maybe more than one. It is my belief: the creation of a galaxy was observed and thus the creation of matter and energy. Following is my understanding: I believe an Eternal Mind, beyond our comprehension, has created everything that has been created. Therefore, the puzzling question about *how* galaxies are created nags at our intellect. Since I have been "out of my body" and have observed the operation of the "Mystical State of Consciousness" (Supernatural consciousness) I know there is consciousness beyond mortal consciousness that is too amazing to describe and I know the physical universe is composed of energies that are different from the energies of the etheric domain. I believe that energies of the etheric realm can exercise control over energies of the physical realm. Furthermore, I believe the creation of galaxies begins with certain swirling motions of the etheric energies.

Seen upon this Earth are swirling energies that impart powerful forces to the environment, such as tornadoes, hurricanes, and the formation of electrons in a spinning generator/alternator as electrons are formed from the cutting across flux lines of electromagnetism. Also, I know that etheric energy can be affected by directed thought. And now reader, I will jump to the end of this explanation by simply stating: all of creation has been brought into existence by the thoughts of God.

After long eons in etheric activity, God set in motion a plan to create a physical creation. This was accomplished by the interaction of two energies. I believe that there are more than two types of energy; however, I only know of two types, these being the etheric and physical. I believe at certain places in space the etheric energies are manipulated to swirl that begins to form into physical substance and when a critical point is reached a tremendous explosion occurs that creates a galaxy. Ah reader, this is not far from the verse, "and let there be light and there was light."

The Kant-Laplace nebular hypothesis theory is now replaced with a reasonable and authentic process. At the point after a great explosion that created what would become a galaxy, the huge fireball assumed a spherical shape, since that shape allows the greatest amount of substance to be held in the smallest shape. And the gigantic fireball began to spin or rotate, and it still rotates. The explosion spewed forth great globs of plasma, and the rotating motion of the newly formed galaxy held the plasma globs in its rotary gyroscopic extended gravity field. The millions of plasma globs also assumed a spheroid shape to conserve their identity and they became what we call stars or suns.

As the central fireball (now a galaxy) rotated, the rotary gyroscopic gravitational extension that I have named the Aust kept the plasma globs in a central region—that is, the galaxy's zero line of latitude. As the galaxy rotated, the globs of plasma were swept along in counterclockwise rotating spiral arms. The stars or suns now began a creation process that had brought them into existence. The suns were also rotating in a counter clockwise motion and they began to spew forth globs of sun matter or plasma. These small fireballs cooled and became planets and a planetary system was created. Every celestial body that is spheroid had a molten beginning; every celestial roamer that is not spherical and is craggy had its origin from collisions of cooled sun matter.

The reason for planets and moons revolving around their superior body has been given, but *why* a galaxy rotates has *not* been given. The reason why a galaxy rotates is the same reason why galaxies were created. Galaxies are not just fireballs rotating in vast space, galaxies and all celestial bodies have an attachment to a mother energy. Some type of mother energy that is composed of Ethons began to swirl and this swirling created physical matter and it exploded into space and eventually became a galaxy. This is a pycnotic creation event. An interaction between a galaxy and its mother energy causes a galaxy to rotate.

## PROPOSITIONS CONCERNING OUR PLANETS AND OUR SOLAR SYSTEM

Planets of our solar system are huge spheroid-shaped energy masses that spin or rotate on an axis while they revolve around the Sun. Each is held in place by the gyroscopic gravitational force of the Sun; each is driven around the Sun by the gyroscopic, rotary, gravity field of the Sun and each planet's revolving rate of movement or speed is determined by the planet's distance from the Sun (Kepler's third law). Each planet radiates a pulling gravity force according to Newton's law of universal attraction. All the planets are forced to revolve in counterclockwise directions around the counterclockwise rotating Sun.

Therefore, if a body in our solar system is roughly spherical, revolves around the Sun in the same direction of the Sun's rotation, and its distance from the Sun cubed equals its period of rotation squared, it's a planet. Welcome back Pluto. However, after a few scientists decided to kick Pluto from the family of planets, there was such an uproar that these self styled Caesars in the scientific community changed their stance and in effect stated: OK, Pluto is a planet, although a *dwarf* planet.

Planets rotate on their axis due to the engaging of their radiated energy waves in the revolving solar gravity energy confines.

These confines that are created by the rotary gravitational field of the rotating Sun enable planets to give birth to or to create planetary orbs. Planets and moons rotate as spinning tops and are tilted on their axis according to the distribution of their land mass. Planets are located in a rotary, gyroscopic, saucer-shaped, gravitational energy extension of the Sun which I have named the Aust. It has three gravitational regions.

The basic revelation in this work that pertains to the solar system is the proof of an Aust and planetary orbs and how they interact. The three gravitational regions that are presented are not supported by any rationale. I can only submit to the reader and the scientific world the structure of the Aust that I perceived as I studied the solar system.

If it is true, then my perception was accurate. If it is not true, what difference does it make? It makes no difference, since this is the structure of the solar system as I perceive it, I submit the three gravitational regions of the Aust. The regions are as follows: 1st or inner vestal region, the 2nd or middle vestal region. and the 3rd or the outer vestal region. mercury is the only planet in the 1st vestal region.

Venus though Neptune are located in the 2nd vestal region and Pluto is located in the 3rd or outer vestal region. Mercury's unique position in the Sun's rotary gravitational field places Mercury in the first gravitational band and most powerful gravity region in the solar system. Mercury also has the unique power to "cut in" in any resonation and superimpose its influence over all other influences.

Planets move in their tubular enclosures that can be considered circular, or a semi flattened doughnut shape. The planets travel from an 'innermost' point or perihelion to the 'outermost' limits of the confines at aphelion. As a planet moves towards aphelion, the Sun propels it to the 'outer' limit of the confining circular tubular enclosure then as it moves towards perihelion, it is propelled and attracted by the Sun to the 'inner limit' of the tubular confine.

Thus planets move in an eccentric motion that can be accurately defined as uneven ovals. The dynamics of the individual tubular confines obeys the same rule for all the different planetary tubular confines. The closer to the Sun the more rapid are the interactions of the particles that constitute the tubular confine. Each tubular confine varies in width, Mercury and Pluto being the widest. The planetary tubular confines are as though they are located in a huge spectrum extending out from the Sun, with the violet end of the spectrum at the Sun.

As a planet moves from perihelion towards aphelion, its rate of motion decreases while it moves in its tubular enclosure. Upon reaching aphelion, the cycle reverses as the planet moves away from aphelion towards perihelion. Thus it quickens its movement as it moves closer to the Sun in its tubular enclosure.

The shape of the uneven oval or its eccentricity is influenced by the gravity force of other planets. Kepler before me revealed that revolving planets have an oval-shaped orbit. Newton's law of gravitation revealed how interplanetary gravitation affects planetary orbits. I wish to give evidence that will correct an erroneous concept beyond the least increment of doubt: uneven ovals are not a proper ellipse. For the sake of honesty—notice I stated *honesty*—may the term *ellipse* be discontinued when referring to planetary orbits and the truthful term uneven ovals be used in its stead.

The six given illustrations leave no doubt, planetary orbits are uneven ovals.

Since the Sun is revolving around the galaxy and taking

the planets with it, the planets are really moving in a spiraling motion within a tubular swirl.

The uneven ovals are illustrated by reducing the Sun to the size of a dot and using two circles to illustrate the outer and inner limits of the orbital confines. This method enables the orbits to be magnified. The orbits of Mercury, Venus, Earth, Earth's Moon, Mars and Jupiter are given. All are tracked by using the United States Naval Observatory data—MICA—version 1.0 Seven of the planets have a nearly uniform shaped orbit while Mercury and Pluto have a very elongated orbit.

The first illustrated uneven oval is for Earth 1996 - 1997. Heliocentric longitudes in a.u.

| | | |
|---|---|---|
| July 5, 1996 | 283 degrees | 1.01671 Aphelion |
| July 15 | 293 | 1.01647 |
| July 25 | 302 | 1.01567 |
| Aug. 3 | 311 | 1.01460 |
| Aug. 16 | 324 | 1.01251 |
| Aug. 24 | 331 | 1.01084 |
| Sept. 4 | 342 | 1.00829 |
| Sept. 13 | 351 | 1.00600 |
| Sept. 23 | 0 | 1.00320 |
| Oct. 5 | 12 | 0.99980 |
| Oct. 13 | 20 | 0.99752 |
| Oct. 23 | 30 | 0.99466 |
| Nov. 4 | 42 | 0.99153 |
| Nov. 12 | 50 | 0.98964 |
| Nov. 22 | 60 | 0.98747 |
| Dec. 2 | 70 | 0.98575 |
| Dec. 22 | 91 | 0.98358 |
| Dec. 25 | 94 | 0.98342 |

| | | |
|---|---|---|
| Jan. 1, 1997 | 101 degrees | 0.98326 |
| Jan. 11 | 111 | 0.98346 Perihelion |
| Jan. 21 | 121 | 0.98409 |
| Jan. 31 | 131 | 0.98528 |
| Feb. 13 | 144 | 0.98745 |
| Feb. 20 | 152 | 0.98886 |
| March 4 | 164 | 0.99174 |
| March 12 | 172 | 0.99383 |
| March 22 | 182 | 0.99655 |
| April 3 | 193 | 1.00004 |
| April 11 | 201 | 1.00233 |
| April 21 | 211 | 1.00507 |
| May 3 | 223 | 1.00826 |
| May 11 | 230 | 1.01016 |
| May 22 | 241 | 1.01241 |
| May 28 | 247 | 1.01351 |
| June 3 | 253 | 1.01447 |
| June 13 | 262 | 1.01565 |
| July 4 | 282 | 1.01675 Aphelion |

Aphelion A.U. is 1.01671 while perihelion A.U. is 0.98326. The difference is 0.03345. That makes the width of the confine about 3,110,850 million miles.

To magnify the orbit, choose 33 increments from aphelion to perihelion. Aphelion and perihelion are now separated by 33 increments. Place aphelion Earth at 283 degrees and 1.01671 units. Place perihelion Earth at 101 degrees and 0.98326 units. All positions must now be graphed in the space between the circular confine.

Below is a magnification of Earth's orbital path according to the US data

Below a larger graph of Earth's uneven oval orbit is a better illustration of the true orbital shape.

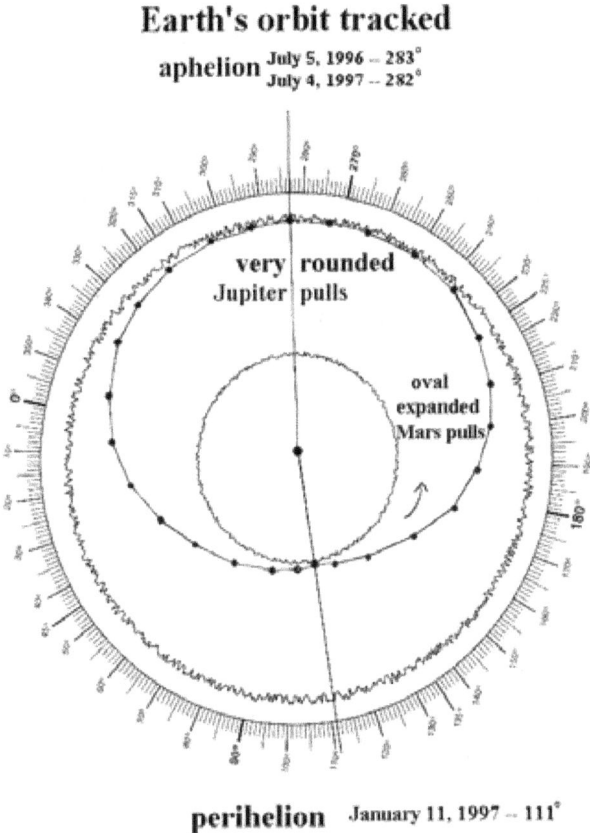

**Earth's orbit tracked**

aphelion July 5, 1996 -- 283°
July 4, 1997 -- 282°

very rounded
Jupiter pulls

oval
expanded
Mars pulls

**perihelion** January 11, 1997 -- 111°

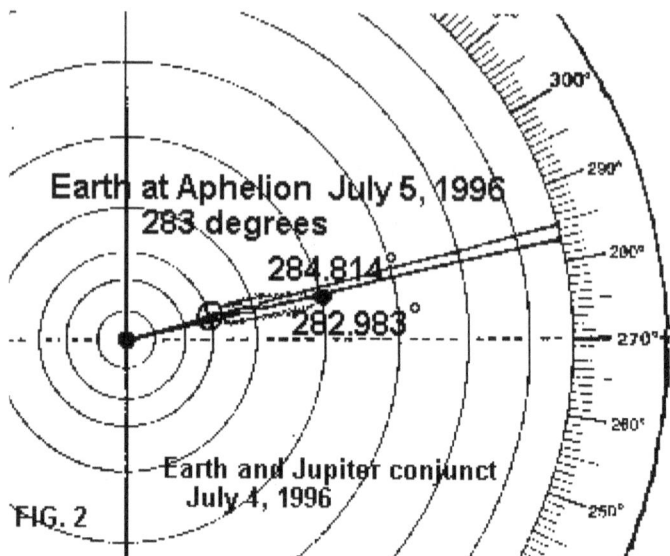

Earth at Aphelion July 5, 1996
283 degrees
284.814°
282.983°

Earth and Jupiter conjunct
July 4, 1996

FIG. 2

300°
290°
280°
270°
260°
250°

Each planetary orbit is shaped differently, dependent upon the shape of the tubular enclosure and the effects of interplanetary gravitation. Each planetary orbit has characteristics of an ellipse since a line drawn from the center of the Sun to the center of the respective planet will illustrate how a planet will sweep through equal areas in equal times. Thus, the quotient from multiplying different areas by the time traversed is always equal.

By using the US government's astronomical data from July 6, 1996 to July 4, 1997 and magnifying the orbit, the Earth's orbit is easily seen to be a perturbed uneven oval. Notice how the orbit is more flattened at perihelion compared to aphelion, and going toward aphelion the oval is more expanded compared to going away from aphelion. Would you care to speculate as to the reasons for these differences? Allow me to explain. It is not a relativistic affect. I have included two drawings that speak graphically to explain the differences. Jupiter's pulling force shown on Figure 2 illustrates how Jupiter pulls on the Earth during the conjunctive period that reached an exact center line on July 4, 1996. For about 15 days before aphelion and 15 days after aphelion, big Jupiter pulls on the Earth and causes Earth's oval orbit to be very rounded, or perturbed, at the aphelion point.

March 17, 1997 the Earth conjuncted Mars. Earth's orbital motion and its oval were altered or perturbed. Earth's uneven oval-shaped orbit is slightly bulged or pushed out toward Mars for more than a month as Mars pulls on the faster moving Earth. See Figure 3, for March 22, 1997.

Every yearly orbital cycle of Earth possesses a different oval-shaped orbit, dependent upon where the Earth passes Mars and Jupiter. The Earth was at aphelion in 1996 on July 5, but supposing over large expanses of time the points of aphelion and perihelion revolve so that some time the point of aphelion would be during December. Also suppose that Earth, Jupiter, and Mars are in conjunction or close proximity at the time of Earth's aphelion. That situation would create a condition where the Earth would be the farthest from the Sun than usual and Jupiter would slow the Earth's orbital movement so that during December and January the Earth would sustain its general position for a longer period. Would you care to speculate what effect would that have on the Earth's weather? Needless to say, it would be a long, cold, severe winter in one latitude and a long, mild, pleasant summer in the opposite latitude. Ancient philosophers knew that certain planetary positions could have a powerful affect upon the Earth's weather. Scientists in my time (about the year 2000) believe that: 1. Changes in the Earth's axis tilt; 2. periodic wobbles in the tilt; and 3. changes in the Earth's elliptical orbit produce ice ages. While the first two may have validity, how could changes in the Earth's elliptical orbit be a factor when Earth's orbit is not elliptical but is an uneven oval?

I now give to you, for your study, the magnification of the orbital paths for Mercury, Venus, Mars, and the Earth's Moon.

Mercury's orbit tracked 1996 United States Naval Observatory data MICA heliocentric longitude of Mercury - distance in A.U.

| Jan. 12, 1996 | 79 degrees | .30752 perihelion |
|---|---|---|
| Jan. 15 | 98 | .31091 |
| Jan. 19 | 122 | .32351 |
| Jan. 21 | 133 | .33245 |
| Jan. 24 | 149 | .34805 |
| Jan. 27 | 163 | .36510 |
| Jan. 31 | 180 | .38819 |
| Feb. 3 | 191 | .40469 |
| Feb. 7 | 205 | .42453 |
| Feb. 11 | 218 | .44109 |
| Feb. 13 | 224 | .44790 |
| Feb. 21 | 247 | .46481 |
| May 23 | 258 | .46668 Aphelion |
| May 27 | 269 | .46644 |
| May 31 | 280 | .45702 |
| June 4 | 292 | .44571 |
| June 9 | 308 | .42602 |
| June 13 | 322 | .40642 |
| June 17 | 337 | .38436 |
| June 21 | 355 | .36121 |
| June 24 | 9 | .34438 |
| June 27 | 25 | .32929 |
| June 30 | 42 | .31728 |

| July 3 | 60 | .30970 |
| July 6 | 79 | .30753 |

The greatest width of Mercury's confine perihelion is 14.8 million miles. Below the uneven oval orbit of Mercury is truly graphically illustrated.

Einstein claimed that Mercury's orbit was an ellipse and it revolved from a relativistic affect. This failure of Einstein to have never mentioned that planetary orbits are uneven ovals and insisted on referring to their orbits as ellipses reveals to serious researchers that Einstein never plotted their orbits and therefore *never knew* the truth. By plotting the actual orbit of Mercury and magnifying it as given in the above chart, the accurate orbit of Mercury can be seen to be an uneven oval.

Mercury's orbit is not an ellipse.

Urban Jean Joseph Leverrier (March 11, 1811–September 23, 1877), a French astronomer, calculated that the perihelion point of Mercury advanced 38 seconds of arc per century. However in 1882, a more accurate figure placed Mercury's advance at 43 seconds per century. The advancing, which leads to the eventual revolving of Mercury's perihelion point, has vexed astronomers. However, the world-famous Dr. John Wheeler, formerly of Princeton University has checked the 43-second advance of Mercury's perihelion point and has discovered that the attraction of Venus to Mercury is responsible for the revolving of Mercury's aphelion and perihelion points. However, the last word on this slight advancement in Mercury's perihelion and aphelion points may not have been said, for the planets do not revolve around the Sun. For since the Sun is moving as it revolves around the center of our galaxy the planets spiral around the Sun.

# Mercury's orbit tracked
**alphelion** May 23, 1996 -- 258 °

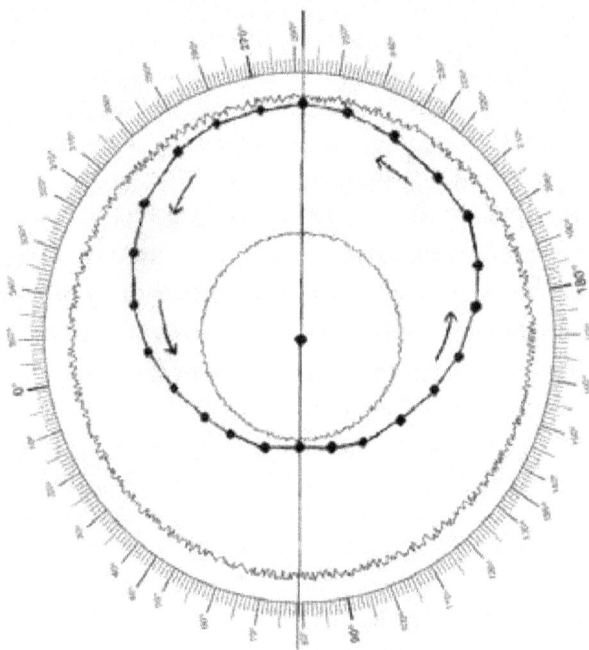

**perihelion** January 12, 1996 -- 79°
July 6, 1996 -- 79°

More conclusive evidence of Mercury's orbit being an uneven oval can be illustrated by further magnifying Mercury's orbit by graphically widening the width of Mercury's confine. When the one and a half inch distance between the small and large circle is increased to two and a quarter inches, the smaller circle becomes smaller, enabling a greater magnification that reveals Mercury seems to collide with the walls of the tubular enclosure in a glancing motion at aphelion and perihelion.

## Mercury's orbit tracked
### Mercury's oval orbit greatly magnified

**aphelion** May 23, 1996 -- 258°

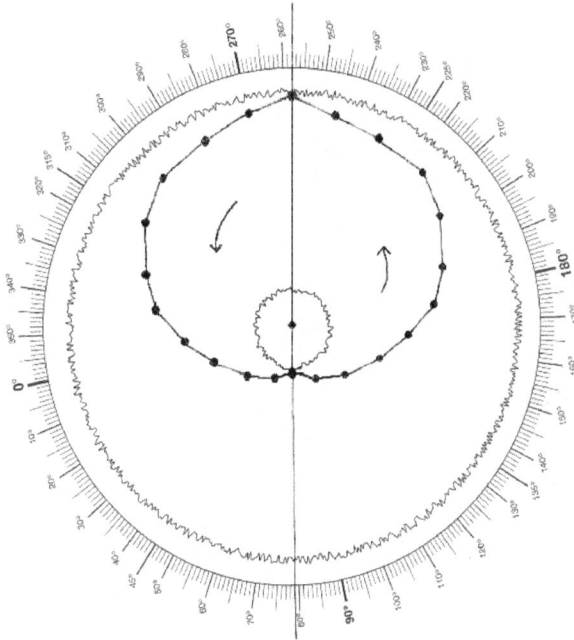

**perihelion** January 12, 1996 -- 79°
July 6, 1996 -- 79°

Venus's Orbit Tracked 1996 United States Naval Observatory Data MICA Heliocentric Longitude Of Venus - Distance In A.U.

| | | |
|---|---|---|
| March 22, 1996 | 131 degrees .71843 perihelion | |
| March 31, " | 145 " .71859 | |
| April 9, " | 160 " .71904 | |
| April 17, " | 173 " .71967 | |

| | | |
|---|---|---|
| April 28 | 191 | .72083 |
| May 2 | 197 | .72132 |
| May 7 | 205 | .72170 |
| May 17 | 221 | .72331 |
| May 21 | 228 | .72385 |
| May 26 | 236 | .72452 |
| June 5 | 252 | .72578 |
| June 17 | 271 | .72704 |
| June 25 | 283 | .72765 |
| June 30 | 291 | .72793 |
| July 9 | 305 | .72821 |
| July 13 | 312 | .72824 aphelion |
| July 17 | 318 | .72820 |
| July 26 | 332 | .72791 |
| Aug. 7 | 351 | .72707 |
| Aug. 15 | 4 | .72628 |
| Aug. 24 | 18 | .72521 |
| Aug. 29 | 26 | .72456 |
| Sept. 3 | 34 | .72389 |
| Sept. 7 | 41 | .72334 |
| Sept. 16 | 55 | .72212 |
| Sept. 19 | 60 | .72170 |
| Sept. 22 | 65 | .72132 |
| Sept. 26 | 71 | .72084 |
| Oct. 7 | 89 | .71967 |
| Oct. 15 | 102 | .71903 |
| Oct. 24 | 116 | .71856 |
| Nov. 2 | 131 | .71840 perihelion |

The width of Venus's confine is not very wide, a little over 915 thousand miles, making its orbit almost circular.

Below is a magnification of Venus's orbital path. I do not have to say that it is not elliptical. It is an uneven oval with characteristics of an ellipse.

## Venus' orbit tracked
### alphelion July 13, 1996 -- 312°

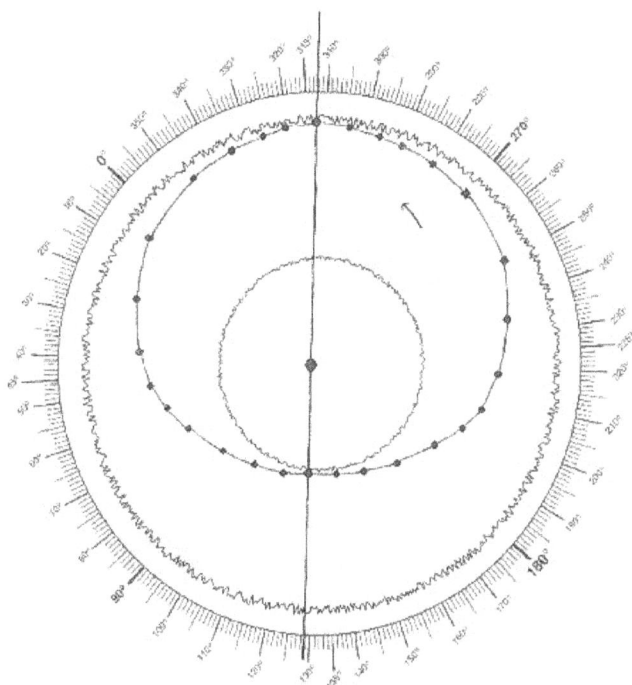

perihelion    March 22, 1996 -- 131°
              November 2, 1996 -- 131°

**LET US FOCUS ON THE UNEVEN OVALS OF MARS AND JUPITER.**
Mars's orbit tracked In 1995, 96, 97 from aphelion to perihelion and returning to aphelion United States Naval Observatory Data MICA For Mars.

Heliocentric longitude distance in A.U.

| | | |
|---|---|---|
| March 13, 1995 | 156 degrees | 1.66598 aphelion |
| April 15 | 170 | 1.66073 |
| May 17 | 184 | 1.64557 |
| June 14 | 197 | 1.62476 |
| July 15 | 211 | 1.59452 |
| Aug. 4 | 226 | 1.55977 |
| Sept. 11 | 240 | 1.52416 |
| Oct. 8 | 254 | 1.48898 |
| Nov. 6 | 271 | 1.45285 |
| Dec. 9 | 290 | 1.41759 |
| Dec. 31 | 304 | 1.39965 |
| Jan. 27, 1996 | 320 | 1.38558 |
| Feb. 20 | 336 | 1.38148 perihelion |
| March 11 | 350 | 1.38500 |
| April 10 | 7 | 1.39911 |
| May 3 | 21 | 1.41779 |
| June 5 | 41 | 1.45308 |
| July 3 | 56 | 1.48791 |
| July 30 | 71 | 1.52306 |
| Aug. 28 | 85 | 1.55992 |
| Sept. 27 | 100 | 1.59467 |
| Nov. 4 | 118 | 1.63060 |
| Nov. 30 | 129 | 1.64855 |
| Dec. 28 | 142 | 1.66100 |
| Jan. 28, 1997 | 155 | 1.66592 aphelion |

Kepler was not the only astronomer that was vexed by
the orbit of Mars. The orbit of the fiery red planet has vexed
observers more than any other planet for centuries. Simply
because Mars's normal 15 miles per second orbital movement
can be seemingly reduced and increased, also due to the Earth
passing Mars it seems (an illusion) to be stationary and move
backward.

Below is a magnification of Mars's orbit. it is an uneven oval
with characteristics of an ellipse.

## Mars' orbit tracked
## aphelion
March 13, 1995 - 156°
January 28, 1997 - 155°

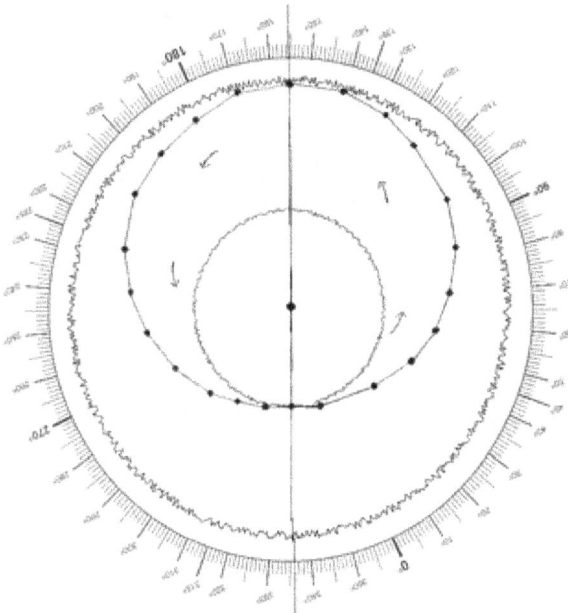

**perihelion** February 20, 1996 -- 336°

Mars' orbit - 1995, 1996, 1997

# "Every planet describes an uneven oval as it maintains the Sun as its focal point in its planetary revolution" -- Johann Kepler

It is not only the planets describe an uneven oval, for the Earth's Moon and other satellites describe an uneven oval as they maintain their focal point in their revolution about their superior body. Yet, the ovals can be seen to be perturbed when a faster moving planet that has sufficient mass to exert gravitational force upon the slower moving planet, as it moves toward conjunction, conjuncts, and then moves away from conjunction. In addition to affecting the shape of the orbit, there is a quickening and slowing of the planets due to gravitational forces. Examples of this effect are given in the following drawings.

It is easily seen how Earth and Jupiter can affect the orbital speed of Mars by a noticeable retarding or accelerating force, dependent upon their position. I submit two drawings to illustrate this effect.

When Mars was at aphelion - 155.866 degrees - on March 13, 1995, Earth was at 172.596 degrees. That placed Earth at about 13 degrees in front of Mars, and capable of pulling on Mars thus quickening its movement.

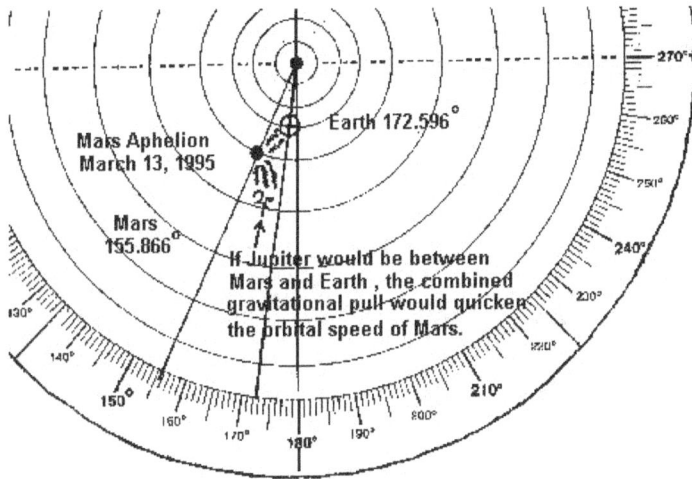

At the next aphelion position for Mars on January 28, 1997, Earth is behind Mars by about 27 degrees and is able to retard the movement of Mars, but if Jupiter was positioned as illustrated, the combined effect of Earth and Jupiter would have greatly retarded Mars's forward movement.

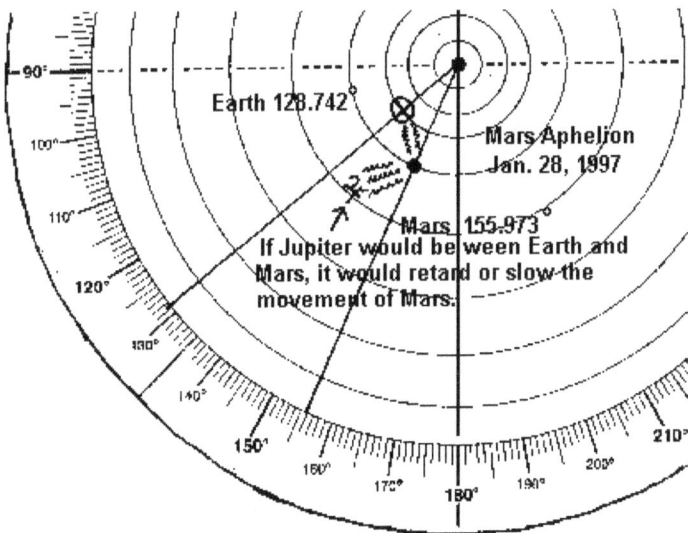

Jupiter's orbit tracked from 1987 to 1999 from perihelion to aphelion and returning to perihelion United States Naval Observatory data for Jupiter MICA from June 14, 1993 Heliocentric longitude distance in A.U. degrees

| July 3, 1987 | 14 degrees | 4.95035 Perihelion |
|---|---|---|
| Nov. 11, 1987 | 24 | 4.95317 |
| Feb. 23, 1988 | 36 | 4.96607 |
| June 1, 1988 | 45 | 4.98183 |
| Oct. 23, 1988 | 58 | 5.01330 |
| Feb. 15, 1989 | 68 | 5.01330 |
| May 28, 1989 | 77 | 5.07625 |
| Nov. 28, 1989 | 93 | 5.13939 |
| May 20, 1990 | 107 | 5.20227 |
| Nov. 12, 1990 | 122 | 5.26547 |
| May 21, 1991 | 137 | 5.32837 |
| Sept. 5, 1991 | 145 | 5.35978 |
| June 1, 1992 | 155 | 5.39127 |
| June 9, 1992 | 167 | 5.42275 |
| Sept. 24, 1992 | 175 | 5.43853 |
| Jan. 27, 1993 | 185 | 5.45036 |
| June 14, 1993 | 195 | 5.454280 aphelion |
| Oct. 16, 1993 | 205 | 5.450620 |
| Feb. 26, 1994 | 214 | 5.438567 |
| June 15, 1994 | 223 | 5.422714 |
| Nov. 17, 1994 | 235 | 5.391400 |
| March 21, 1995 | 244 | 5.359741 |
| July 6, 1995 | 253 | 5.328360 |
| Jan. 12, 1996 | 268 | 5.265447 |
| July 6, 1996 | 282 | 5.202256 |
| Dec. 26, 1996 | 297 | 5.139322 |

| | | |
|---|---|---|
| June 27, 1997 | 313 | 5.076451 |
| Oct. 7, 1997 | 322 | 5.044830 |
| Jan. 30, 1998 | 332 | 5.013382 |
| June 22, 1998 | 345 | 4.982053 |
| Sept. 29, 1998 | 354 | 4.966232 |
| Jan. 13, 1999 | 4 | 4.955194 |
| May 18, 1999 | 15 | 4.950467 perihelion |

Below is a magnification of Jupiter's uneven oval its orbit is an uneven oval with characteristics of an ellipse

## Jupiter's orbit tracked

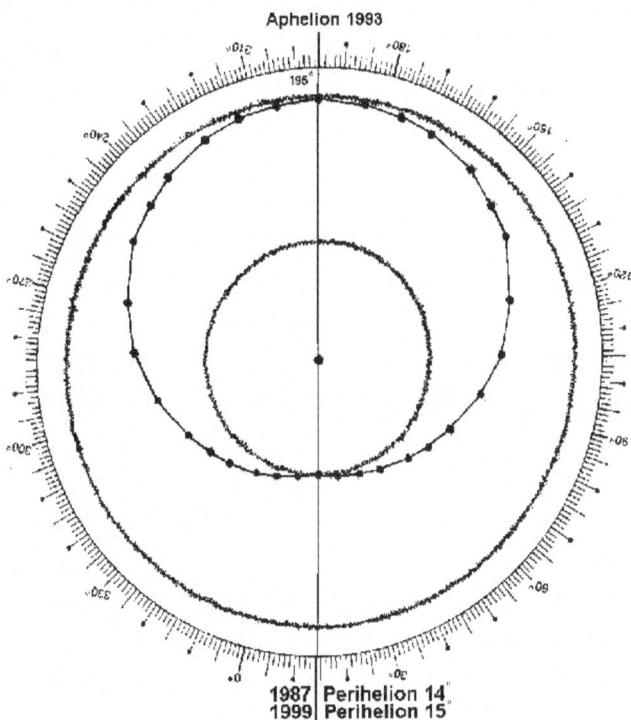

Aphelion 1993

1987 | Perihelion 14
1999 | Perihelion 15

The revelation that planetary orbits are not a proper ellipse does not detract from the importance of Kepler's work, for this work proves Kepler's work.

Kepler originally claimed the orbits were uneven ovals; however, due to a sure refusal of astronomers to accept the concept of uneven ovals over the concept of circles, Kepler used "political correctness" because of the entrenched false concepts and chose to use the term for the orbital shape as an ellipse. This was an easy escape route for Kepler, because the planet's uneven oval orbits obey the law of an ellipse.

Yet, the works of Copernicus, Galileo, and Einstein have been immortalized in the annals of science by having statues built in honor of their tremendous contributions, but only one statue, in Prague, has been erected to honor the great Kepler and Tycho Brahe.

The university in the city of Linz where Kepler taught is long gone but about 1970 a new liberal arts college was built in Linz and named the Kepler College to commemorate the great accomplishments of Kepler. But Kepler was an astronomer, mathematician, and taught these subjects where he made his debut into the world of science, at the university at Graz. It is recorded that Kepler was the last great renowned scientist to grapple with astrology secrets, although there probably were other unknown researchers that grappled with these secrets of the solar system.

We do know that Kepler was ridiculed for his beliefs and felt great embarrassment. However, Kepler was an assistant to Tycho Brahe and Tycho was the astronomer for the Holy Roman Empire. After Tycho's demise, Kepler was appointed to the position of Imperial Mathematicus of the Holy Roman

Empire. Therefore, Kepler not only was an advanced astronomer, he had a prestigious position; that was a factor in his ability to endure the ridicule. He was resolute about his comprehensions of the solar system, and he believed that his work would be honored in future times.

### KEPLER'S THREE LAWS

Kepler's first law states: every planet follows an elliptical path around the Sun with the Sun as its focal point.

Kepler's second law states: an imaginary line from the Sun to a planet sweeps equal areas in equal times. This law also applies to the motion of the satellites of planets or a planet's moons.

This can be illustrated by showing a magnified orbit of our Moon as it revolves about the Earth. An improved method upon the daily or 24-hour data as is given here would be to make a chart based upon readings taken every eight or twelve hours. But the chart given here is ample to show the uneven oval lunar orbit that does obey Kepler's second law.

Moon's geocentric longitude is calculated for 12:00 noon EST beginning January 5, 1996 when the Moon is at apogee and ending February 1, 1996 when the Moon returns to apogee.

Moon's Longitude

| Jan. 5 | 103.054 | .00271 apogee |
|--------|---------|---------------|
| Jan. 6 | 116.231 | .00271 |
| Jan. 7 | 129.487 | .00271 |
| Jan. 8 | 142.583 | .00270 |
| Jan. 9 | 155.760 | .00268 |
| Jan. 10 | 168.936 | .00266 |
| Jan, 11 | 182.113 | .00263 |

| | | |
|---|---|---|
| Jan. 12 | 195.289 | .00260 |
| Jan. 13 | 208.465 | .00256 |
| Jan. 14 | 221.642 | .00252 |
| Jan. 15 | 234.818 | .00248 |
| Jan. 16 | 247.995 | .00245 |
| Jan. 17 | 261.171 | .00242 |
| Jan. 18 | 274.347 | .00240 |
| Jan. 19 | 287.524 | .00239 perigee |
| Jan. 20 | 300.700 | .00240 |
| | | |
| Jan. 21 | 313.877 | .00241 |
| Jan. 22 | 327.053 | .00244 |
| Jan. 23 | 340.229 | .00247 |
| Jan. 24 | 353.406 | .00251 |
| Jan. 25 | 6.582 | .00255 |
| Jan. 26 | 19.759 | .00260 |
| Jan. 27 | 32.935 | .00263 |
| Jan. 28 | 46.111 | .00267 |
| Jan. 29 | 59.288 | .00269 |
| Jan. 30 | 72.464 | .00270 |
| Jan. 31 | | .00271 |
| Feb. 1 | 98.817 | .00271 apogee |

A magnified orbit of the Moon is also useful to show its uneven oval shape, but this is for in-depth analysis because the Moon's orbit is almost circular. Our Moon revolves in the orbital enclosure—that is, the Aust that is created by the rotating Earth—and its revolving has the same characteristics as the Sun's revolving planets. The Earth's Aust has a comparative shape as the Aust for all the planets and that being a shape of a slightly flattened doughnut or donut if you like the modern spelling. The Moon revolves to the outer and inner extremes of its circular confine (perigee and apogee) and to the upper and lower extremes in its circular confine. Thus the Moon's position is measured by longitude and latitude.

The Moon's orbit is slanted only 5 degrees to the ecliptic and its orbital confine varies only about 30,000 miles in its total width. Although, as a microscope magnifies and reveals great details, a magnification of the Moon's orbit reveals its orbit to be an uneven oval with a characteristic anomaly at perigee. This anomaly is true for planets and satellites as they move in their orbits to the closest point to their controlling body.

The Moon's orbit is slowly changing, which is not true for every satellite in our solar system. As the Moon moves in its orbit, it passes Mars and Jupiter once every revolution and at times as the Moon moves it passes Mars and Jupiter while they are in conjunction. This tremendous pull of Mars and Jupiter (earthquake alignment) together pulls the Moon slightly away from the Earth at an average rate of three inches per year.

Although, the distance was verified by NASA and the reason is my idea, I have never read anywhere *why* the Moon is moving away from the Earth in any of my studies, and I believe this concept is true, and is new with this work. I tracked the orbit of the Moon one month and I submit it to you for your enlightenment and amazement. If Isaac Newton had seen this, he would have experienced euphoria.

**Moon's orbit tracked for one month cycle**
apogee points  January 6, 1996 and February 1, 1996
longitudes  103.064  and  98.817
apogee

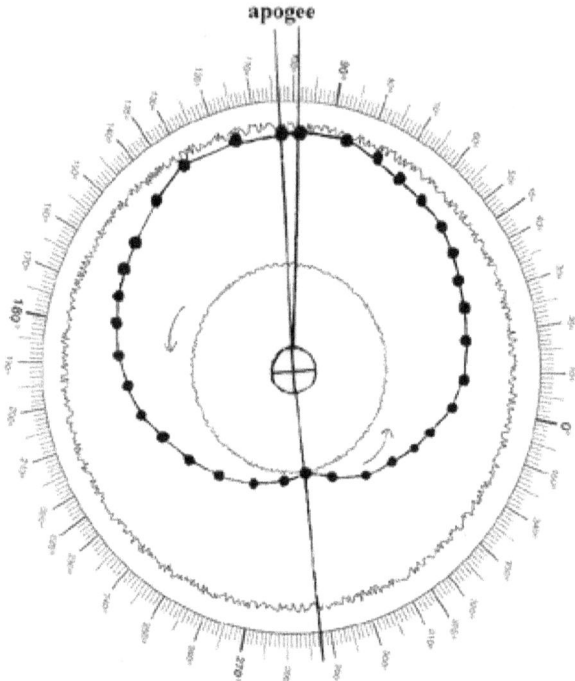

perigee  January 19, 1996
longitude 287.524

Thus, the Moon's orbit magnified is useful also to show the continuity or common pattern of movement for *all* subordinate bodies to rotating superior bodies. Kepler's 2nd law is not a law for our local space but is a law for the entire universe.

You have now seen the orbital shapes of Mercury, Venus, Earth, Mars, Jupiter, and the orbit of the Earth's Moon.

**LET'S CONSIDER THE DEFINITIONS FOR TWO NEW TERMS, IN MORE DETAIL.**

AUST and WAKE.

All rotating spheres have an Aust. It is important to understand this because a planet's orbs are formed within its Aust.

*Aust*

The word Aust has been coined from the word Auschwitz. From the perspective of a newly coined word, it is designed to memorialize those that became victims of the Hitler regime. Let us never forget why and how this word was coined. I was guided by the Holy Spirit to create the word Aust from one of Hitler's most notorious murder camps, where torture preceded murder. Aust, therefore, will memorialize those that died in the Holocaust, with special respect to the brutally murdered innocent children.

*The Nature and Dimensions of an Aust*

An Aust is an extension of rotary gravitational forces radiated from the central girths greatest diameter of a rotating celestial body, where the central girth is 90 degrees to its axis. The size, mass and rotational movement of each celestial body determines the characteristics of its Aust. Our Sun's rotation creates a rotary gravitational force or Aust that is extended outward from its area of greatest girth.

An additional factor adding strength and intensity to the Aust is the amount or "degree of bulge" at its central girth due to centrifugal force. The bulge at the central girth of *all rotating celestial spheres* is the result of centrifugal force.

The bulge is not the cause or reason for the creation of an Aust, the bulge only strengthens and intensifies the Aust. More graphics are submitted to make your understanding of the Aust more lucid. At this place, a graphic of Saturn's rings enables all readers to grasp the idea of a powerful extended gravity field that has power to keep particles of mass locked in its extended rotary gravity field.

**Saturn's Aust keeps its rings in place**

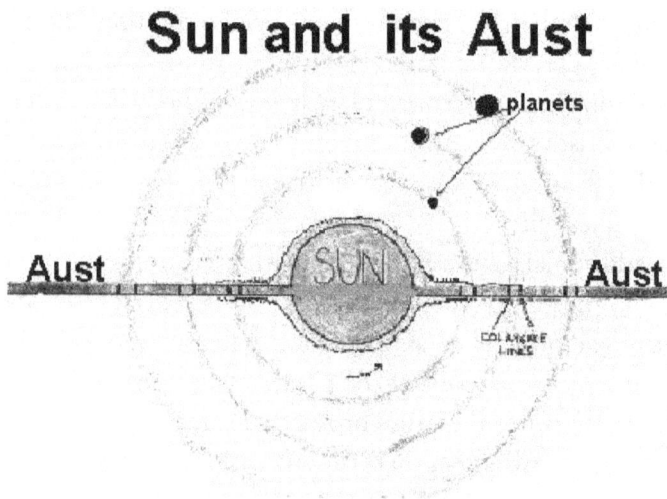

**Sun and its Aust**

planets

Aust                    Aust

Above: two views—top and side—shown together.

You can think of the Aust as the force field radiated by a gyroscope. What is the accepted explanation that keeps a gyroscope fixed in its position? Please consider my analysis and explanation.

A gyroscope does not maintain its position due to a relativistic affect. If a gyroscope be magnetized, and forced to move around a central point in a 30-foot diameter circle, the spinning gyro will not maintain its position due to a inherent power of matter to "warp and deform" space in a "space time continuum."

A gyroscope maintains an inertialized position from its steady angular momentum. What exactly is steady angular momentum? Steady angular momentum is: "1.- Uniform Rotary Movement, - 2. Plane Polarized, - 3. Gravity Field, 4. - of its mass." If any one of the four conditions is altered, a gyroscope will change or lose its inertialized force.

We take for granted the stabilized position of a gyroscope while it is spinning at its maximum rotational rate on this Earth, but as Neil Armstrong demonstrated Newton's law while jumping on the Moon, the movement of mass is dependent upon the gravity field in which it moves. Thus a gyroscope would maintain a "Plane Polarized" position anywhere in the universe that it operates, but its resistance to any angular changes—that is, the force required to move it from its "Plane Polarized" position—will differ according to the gravity field in which it operates. In outer space, where gravity is so weak it can be considered nonexistent, a gyroscope will still operate because the spinning wheel has mass, but very little force would be required to move it from its "Plane Polarized" position.

*A Review of a Gyroscope's Nature.*
A gyroscope is a physically energized "Plane Polarized" spinning device. Its amount of dynamic energy or kinetic energy is directly proportional to its mass and its rate of movement. Therefore, the stored energy depletes slowly as the wheel turns due to friction and gravity pulling on the spinning wheel.

Nutation or wobble of a spinning wheel or sphere: when the spin of a gyro wheel or sphere develops sufficient centrifugal force of its mass to stabilize it in a "plane polarized" position, the spinning object is in a stable inertialized position. It will remain in its inertialized position as long as its gravitational, rotary "Plane Polarized" force is sufficient to maintain a stable position. However, if the gyro wheel or sphere loses some rotation motion—that is, if it slows due to interplanetary gravitational forces, it will lose some of its stable "Plane Polarized" position. In addition and more importantly, if there is a change in the distribution of its land mass, equilibrium forces always adjusts its gyroscopic inertialized force. A wobble or nutation will occur when the stable position weakens.

I submit: Uranus experienced this same situation.

Uranus began to wobble and finally tilted to a 98-degree angle from a redistribution of its mass. (so sez me Jack R. Truett Sr.) This probably happened about a billion years ago or so. Its axis is now pointed directly to the Sun as it rotates in the Aust of the Sun. Its rings and 21 moons revolve in an up–and–down position in the Aust of Uranus, as Uranus revolves in the Aust of the Sun.

Aust: is a gyroscopic gravitational force field that extends outward from a rotating sphere. As a sphere with a bulged girth rotates, innumerable points, on different latitudes, of the sphere, pass through different distances at the same time. This condition creates an Aust. The bulged area at the central girth is traveling much faster than any other area of the sphere.

Since the circumferential distance of the lines of latitudes slightly above and slightly below the midpoint of the girth is slightly less, a non-uniformity of gravity field at this area creates a perturbed gravity belt or Aust. An Aust is formed by the uneven rate of movement of gravity field just above and below the central girth. The gravity field radiated at the area of greatest girth has the fastest rate of movement since this region travels the greatest circumferential distance.

Directly above and below the central girth, the circumferential distance begins to reduce, and this reduced distance causes the radiated field to be slower in movement since this area of the rotating body passes through less distance in the same time.

The uneven movement of gravity field above and below the central girth creates a type of gravity vortex or eddy currents that I have named the Wake. These unequal and uneven forces interact with each other, and the interplay of these uneven forces creates a sandwiching affect where a turbulence created by this unevenness sets an extended gravitational belt that is further influenced by eddy currents due to the turbulence. The turbulence and eddy currents set in motion by the rotating celestial sphere, creates tubular vortices that are planetary orbits. On Earth, the gravitational eddy currents above and below the rotating Earth's Aust causes hurricanes. More about this later at earthquakes, volcanoes and hurricanes.

Further illustration of the Aust can be given by a comparison to air and water. The unequal heating and cooling of the Earth creates air currents. As air currents of differing rates of movement pass each other eddy currents are created. Hurricanes and tornadoes are created by this eddy current affect. With hurricanes, the motion of the rotating Earth causes the clockwise and counterclockwise gravitational eddy current force above and below the Earth's Aust, to interact with the atmosphere. When this occurs, the atmospheric pressure from warm air is diminished to some degree and air rushes in to fill that swirling slight depression and the "rushing in air" in northern latitudes begins to swirl in a counter clock wise direction. The gravity force and the atmosphere are both turning above the Equator at a clockwise direction.

Hurricanes are (low pressure) tropical depressions and are erroneously called—tropical cyclones—this reference is misleading for a hurricane has characteristics to a—dust devil—since both depend on great temperature differences between the atmosphere and the surface of the earth. However cyclones and dust devils are creatures from two fast moving air currents that pass each other, whereas tropical depressions or hurricanes above the equator, begin to rotate from the initial force of "air rushing in" to fill a low pressure depression, then are given a swirling motion by the Wake.

In a hurricane's early state of creation, the low-pressure depression is the eye of the hurricane. As surrounding air rushes in to fill the low-pressure depression, the rushing-in air is *in* the Earth's Aust. Therefore, the rushing-in air is driven by the Earth's gravity field. Thus the rotating air mass—that is, the hurricane—resembles an electric motor. The rotating gravity-driven air can be compared to the field of an electric motor and the eye can be compared to an armature of a motor. The Earth's swirling gravity field in the Wake of the Aust imparts a rotating life to the hurricane and its low pressure imparts a cohesive force to maintain stability to its structure. If either of these two conditions were to be somehow nullified, the hurricane would become a tropical storm. For example, a hurricane cannot exist in the waters of the equator where the Earth's Aust is not swirling in a clockwise or counterclockwise direction, and, if a high-pressure area would push against a low-pressure hurricane, the hurricane would move in the opposite direction to the high pressure force.

*Tornadoes*

Most of the tornadoes in America are formed in "tornado alley" —that is, the flat range area between the Rocky Mountain range and the Appalachian Mountain range. This flat range area offers an "alley" or opening for fast-moving, moist, hot air from the south rushing north and meets fast-moving dry, cool air from the north moving south. When the two fast-moving air currents pass each other, at their region of contact a great air disturbance is created which causes a swirling eddy current, and when this eddy current is seen as a funnel shape from the clouds to the ground, it is known as a tornado. Depending upon the meeting angle of contact of these fast-moving air currents, a swirling of air can rotate in a counterclockwise motion or a clockwise direction above the equator. However, hurricanes, tornadoes, cyclones, and dust devils are influenced by the force of gravity.

Although cyclones and dust devils are reported to swirl in counterclockwise movement above the equator, the same as many tornadoes, the point of origin for cyclones and dust devils *is not* in clouds far above the Earth. A tornado's distinguishing characteristic is a funnel. Cyclones, dust devils, and hurricanes do not have a funnel. Tornadoes are formed by fast-moving air currents that pass each other far above the surface of the Earth. Cyclones are formed by fast-moving air currents on or near the surface of the Earth. Dust devils are formed by a similar force where great temperature differences prevail, usually on the plains or the desert.

In the case of flowing water currents of differing rates of movement, when they come into contact, turbulence is generated and eddy currents or whirlpools are created. If the differing rates of movement increases, correspondingly, great turbulence is generated. Fast-flowing water around a rock in a river or stream causes the water that flows around the rock to slow as it contacts the rock, and as the slower-moving water that flows around the rock meets faster-moving water on the downstream side of the rock, a whirlpool or eddy current is produced. Thus, take very careful notice: when a faster-moving fluid (either liquid or air) passes a slowing-moving fluid, a whirlpool or eddy current is produced.

With an airplane, due to the shape of the wing, air is caused to move slower as it flows across the front of the wing, and as it passes the wing surface the air moves faster and at the point where the slowing-moving air begins to move faster, a low-pressure area is created. This low-pressure area or semi-vacuum pulls the plane upward and this gives the plane lift.

Therefore, in each of these cases given, the principle discovered by the Dutch/Swiss physicist/ mathematician Daniel Bernoulli (1700–1782) is in operation. Bernoulli's law states: the speed of a fluid varies inversely with pressure; an increase of speed produces a decrease in pressure. This law also applies to gravity waves or a gravity force so that not only will the mediums of water and air be affected when two streams of a fluid traveling at different rates of movement pass each other, the interaction from differing rates of movement of rotary gravity fields also creates gravitational disturbances or gravitational eddy currents.

If the Sun was shaped as a cylinder, all gravity waves would be radiated at the same rate of movement, thus an Aust could not be created. Rotating large spherical bodies in the universe creates large rotary gravitational disturbances or a large Aust. Thus, once you get the idea understood, it's easy to grasp the idea that an Aust is an extended gyroscopic rotary gravitational force field. You can get the understanding of the Aust by comparing it to the funnel of a tornado that is laid in a circle around the Sun.

This principle is explained again in greater detail. As the Sun rotates, the differing circumferential distances in its lines of latitude remain about the same. The Sun's circumference at its greatest girth or zero line of latitude has been measured to be about 2,700,000 miles. The rotation of the Sun, though varying at its different latitudes, remains constant. Its rotational speed at its zero line of latitude (equator region) is the fastest, being measured to be about 4,500 mph. This region (zero line of latitude) rotates once in slightly more than 24 Earth days; therefore, the turbulence or perturbed regions at its central girth remains constant. The result of the interaction of gravity energy moving at different rates of movement just above and just below the central girth as the central girth is extended from the Sun creates an Aust and creates tubular vortices that are shaped as a semi-flattened doughnut or a funnel of a tornado laid in a circle around the Sun. These tubular vortices that are planetary orbits have different widths and depths.

So readers, Einstein was getting close to the reason why planets revolve as they do and are not pulled into the Sun. However, Einstein's idea of a trough that was supposed to be the path for planets to use to circle the Sun was submitted without any rationale, except to say that he believed that space is warped and deformed in the vicinity of matter. Einstein's idea was an unfounded hypothesis, and when the idea for a motional force was attributed to inertia, his concept of solar system dynamics was completely false.

### THE TUBULAR CONFINES THAT PLANETS USE TO CIRCLE THE SUN MUST BE INVESTIGATED.

A space probe could be used to explore the orbits of Mercury and Venus to determine if any gravitational differences exist as the probe enters their tubular confines. I believe that if a space probe is directed to make a 90 percent cut across an orbital confine, it should register a gravitational anomaly.

The tubular orbital confine of Pluto has the greatest width and depth; hence, Pluto has the greatest eccentricity and a 17-degree variance, up and down, from the ecliptic plane. The ecliptic plane is an imaginary straight line that extends outward from the Sun's equator region. Though imaginary, the ecliptic plane is necessary to be able to plot the latitude of each planet.

The Earth also has an imaginary line that is extended outward from the equator. It is called the "celestial equator," and since the Earth is tilted 23.5 degrees on its axis, the celestial equator is tilted 23.5 degrees. The Earth's imaginary line or celestial equator is important to accurately determine the time of the vernal and autumnal equinox. These equal days and nights all over the world, or equinoxes, are determined by calculating when the Earth's celestial equator intersects or crosses the ecliptic plane. Hello spring, hello autumn.

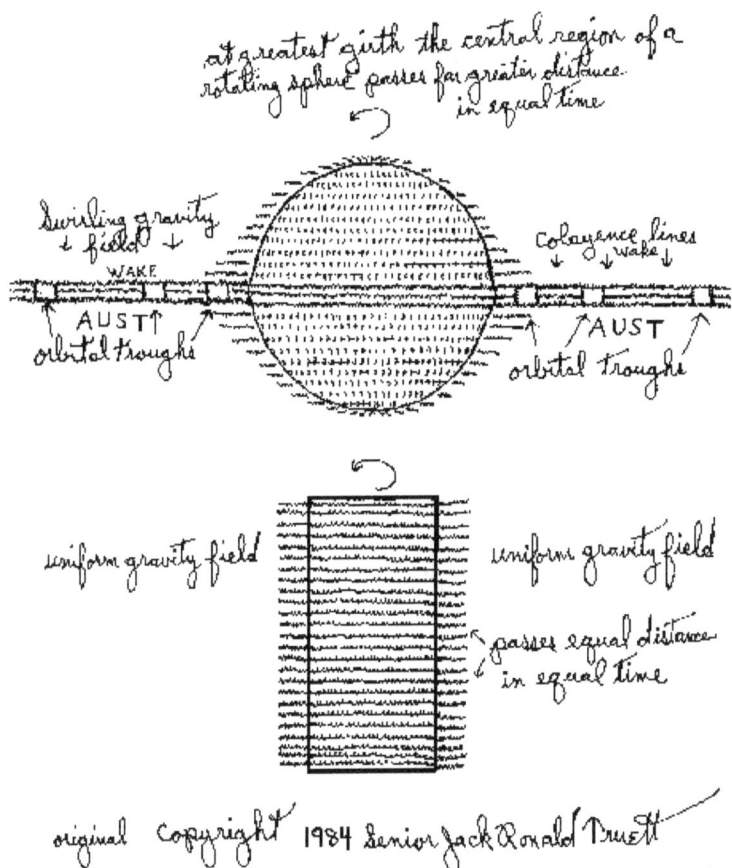

at greatest girth the central region of a rotating sphere passes far greater distance in equal time

Swirling gravity field

WAKE

AUST orbital troughs

Coherence lines wake

AUST orbital troughs

uniform gravity field

uniform gravity field

passes equal distance in equal time

original copyright 1984 Senior Jack Ronald Truett

*Wake*

The uneven movement of gravity field above and below the central girth creates a type of gravity vortex or eddy currents that I have named the Wake. A comparison can be given to the uneven motion of water when a speeding boat is propelled through the water. On both sides of the spinning propeller, a *wake* is formed. In this comparison, the mediums are different, as is the geometry. However, the creation of two different patterns, the trough and the wakes, should enable all readers that try to understand the very natural creation of an Aust and a Wake on a rotating celestial body.

Wake, from the perspective of a new term has also been created for the sciences of physics and astrophysics, from guidance of the Holy Spirit. The term *Wake* is a natural, and it will also memorialize the little children that died at the Waco massacre.

I entreat all that cherish freedom and security and love little children to never forget the motivation that came to my mind to name the gravitational disturbances directly above and below the central girth of a rotating celestial body and to memorialize the murdered little children. The United States government atrocities at Waco, Texas in 1993 that was engineered by Justice Department people, then headed by Janet Reno and approved by Chief Executive Bill Clinton must never be forgotten. The spiritual aspect of the term Wake was created to forever memorialize the 25 little victim children that were brutally murdered by a Masada-type death, and two being crushed by a United States Tank.

The Masada-type death came as a direct result from the trauma created in the minds of those inside the compound by government officials intent upon using Gestapo-type tactics as demonstrated by pumping the deadly CS gas into the compound for six hours and (later proven) illegally used incendiary devices to set fire to the Waco compound, and when two children and their mother were crushed to death by a government tank. Later, during the G. W. Bush presidency, similar compounds that were headed by a Mormon by the name of Warren Jeffs were not bulldozed and not burned to the ground.

Also to memorialize those that died unnecessarily when

their human rights and constitutional rights were violated in Gestapo-type intent to commit murder and actual premeditated murder, for when the tank began to crush all humans into a compacted pile of rubble, it was not known that children were still alive in the compound.

As the atrocities at Auschwitz, Waco was also a careful, premeditated scheme by people in the government. However, in America, the atrocities at Waco was contrived by our own Justice Department and approved by the American Chief Executive, Bill Clinton.

In Hitler's murder camps the SS guards tried to destroy evidence and cover up their deeds as the American forces advanced, and in Waco a more elaborate cover-up was completed as the remaining structure with the dead bodies was bull-dozed into a hole, just as the dead bodies were bull-dozed into holes and ditches in Hitler's murder camps.

If any reader believes that the murdered children at the Waco compound should not be remembered or Bill Clinton and members of his administration could not be as brutal, ruthless, and black-hearted as I am stating, consider the following events that are a known part of history. During Chief Executive Clinton's second term, a now famous or notorious female aide (depending upon your viewpoint) was committing oral sex upon Bill Clinton while he was planning more death and destruction to a foreign nation. For greater clarification: Bill Clinton was talking to a senator about bombing Bosnia, while his female aide had his genitals in her mouth. If history books will not record this terrible act, this book will.

The nature of energy, ethons, electrons, electricity, plus the Aust and Wake that is created by our rotating Sun, has been carefully explained. Before we continue, it is important to clarify the fact of no actual practical value of Einstein's theory of general relativity from another perspective. NASA's space program, and in truth the space program of every nation, is dependent upon very accurate mathematics to determine the escape velocity for rockets leaving the Earth and the gravitational force that determines the distance from the Earth needed to enable any manmade satellite to be placed in orbit. To accomplish these delicate feats, the Newton-Cavendish formula of universal gravitational attraction is necessarily used, and I emphasize, not a least bit of general relativity.

In truth, as in the cases of the atomic bomb and the space program, if Einstein had never been born, the atomic bomb and the space program would have developed, for general relativity was not used in any way to develop the atomic bomb, nor is it used in calculations to perform the delicate feats of the space program. Einstein's theory of general relativity is a false theory, and IF it were true, it would have NO practical use.

Since the useless value of general relativity is made clear, let us review again some of the concepts embedded in general relativity. The three following factual concepts are foundation ideas that Einstein used to formulate his general theory of relativity. These concepts also gave his theory its name. Einstein as Newton both paraphrased an Ancient Roman saying, that each stood on the shoulders of giants. In my unbiased and honest searches of Einstein's work, I never found where Einstein mentioned the name of the Italian roaming philosopher, Giordano Bruno (1548–1600). As a professor at the prestigious Berlin Technical Institute, and certainly attending symposiums, it is unlikely that someone did not mention the very unique idea postulated by Bruno.

Bruno was the first theoretical analyst to correctly state in his writings the following: "Everywhere there is incessant relative change in position throughout the universe, and the observer is always at the center of things." This quote was posted on the Web in a work about Giordano Bruno by John Kessler PhD

This relative principle is a law of universe motion, and Bruno should be given credit for being the first to state this law. Bruno did not mention the time of the observation, for it was understood or it went without saying. Therefore, a paraphrase to Bruno's law of universe motion that includes time is: the place of any celestial body in space is dependent upon the time of the observation and the position of the observer, that is, the *relative position of the observer to the observation point* and when the observation is being made.

**WE CONTINUE WITH EINSTEIN'S POSTULATES.**

I. A long-held geometric axiom was "the shortest distance between two points is a straight line." Einstein proved this to be true only for flat surfaces. He corrected our thinking by proving that "the shortest distance between two points on a convexed, curved surface is a curved line and the shortest distance from point A to point B following a curved surface if the curved surface is moving is also a curved line. However, in this case, rates of movement (and if steady or not) and the shape of the curved surface are factors to be included in determining the shape of the curve from point A to point B."

II. Secondly, stars are not where we see them because we only see their light. It took millions of years for the light to reach the Earth and in that time they moved or ceased to exist. Also, the Earth is moving, that is, our observation point is changing, therefore what we see is dependent upon conditions (i.e., the velocity, the direction of movement and whether these be in a steady state or changing) of the observation point.

Therefore, when looking into outer space we see an *apparent* reality NOT real time reality. This is a repeat of Bruno's Law of Universe Motion.

III. True Concept Introduced: When calculating the rate of movement of moving objects, velocities cannot be added that would exceed the velocity of light or otherwise we would see events happening before they occur. If the Grecian developers of geometry would have used the term 'direct line' instead of 'straight line' as the shortest distance between two points it would not have become a bone to pick by Einstein, for the shortest distance between any two points is a direct line.

If a dog takes a changing path to catch a rabbit that is running in a zigzag course, the dog may have to take a few diagonals but if the dog adjusts his path to keep in a direct line of vision to the rabbit, the dog's path is a direct line, and is the shortest distance between the dog and the rabbit.

The shortest distance up and over a mountain and down the other side is not over the mountain, but is a direct line' through the mountain. The direct line for a traveler on the spherical Earth going from point A to point B may be curved, but if the traveler is traveling by air, the direction and rates of movement are vital factors and the direct line could be a curve, but dependent upon variables as diminishing circumferential measurement and changing rates of movement, the direct line could be an adjusted curve.

For the sincere students that want to burn in their minds the facts about general relativity and want to remember beyond the shadow of any doubt, why Einstein's theory of general relativity is a false theory, I will repeat again the following theoretical concepts woven into general relativity that are false.

We have already shown that the idea of relative motion was first introduced by, Bruno, the Italian roaming philosopher that was burned at the stake in 1600 by the Catholic Church for teaching his concepts.

I. Einstein claimed that a rapidly accelerated body will shrink in its direction of movement, gaining mass as it accelerates, reaching a point of infinitude at the 99th plus velocity of light, where time will stand still. This idea was taken from the theoretical concept known as the Lorentz -FitzGerald Contraction, and has no rationale to explain this supposed phenomena. Neither does it submit any mathematics to prove this idea. The idea is a false as a $3 bill.

A soft electron will gain mass to preserve its identity when greatly accelerated through a magnetic or an electromagnetic field as through magnetic eddy currents in a cyclotron. Einstein claimed that planets revolve around the Sun by inertia. We previously have made this point, because the inertia idea as the driving force for the planet's revolving motion must be completely dismissed.

In logical sequence, the existence and nature of the Aust and Wake must be submitted, before the existence of astrological orbs is submitted. Orbs are created in the tubular confine of each planet, and it is the interaction of planetary orbs that affect human thinking and behavior. When you read Volume II of *My Search for Truth*, you will be prepared to understand how the orbs are created.

II. Einstein claimed that gravity is not an attractive force that pulls all matter towards its center but is a condition of space in the vicinity of matter. Einstein's idea is false because by direct measurement, gravity is proven to be a force that attracts all bodies towards its center.

III. Einstein claimed that the Ether does not exist. Einstein was grossly incorrect for the Ether *does* exist and its particles I submit to be named Ethons. The mirrored box that I built where light energy is dissolved proves that light energy does dissolve. Therefore, it does dissolve into something, and the pycnosis theory offers the sensible explanation as to where light energy dissolves.

IV. Einstein's claim that energy cannot be created or destroyed, but only transformed is false, for my experiments prove that energy can be dissipated and dependent upon conditions, electromagnetic energy can dissolve. When electromagnetic energy dissolves, it dissolves into Ethons.

V. Einstein claimed that time does not "flow evenly" in the universe. This is the concept that has confounded minds and has mistakenly fueled science fiction writers more than any other relativistic concept in Einstein's theory of general relativity.

Atomic and molecular time does change as its thermal motion changes. Any advanced biological life that may travel at great speeds in a space ship will not have great biological changes to their system or they would die from the shock. Therefore, remaining in Einstein's theory of general relativity is the law of universe motion, first submitted by the roaming Italian philosopher, Bruno, who was burned at the stake in 1600, for teaching his concepts and the limitations imposed by the velocity of light.

The two true postulates of Einstein follow: (1) no physical

matter can exceed the velocity of light. And (2), when anyone looks into the starry heavens, the cosmic bodies positions are not exactly as you see them, since we are separated by great distances, we see cosmic bodies as they existed at some time in the past. During my research, I had cause to wonder many times about Einstein's statement: "no amount of experimentation can ever prove me correct, but a single experiment at any time can prove me wrong." I had very good reasons to wonder. Did Einstein know that his piecing together the unrelated theoretical concepts of other theoreticians into a general theory was a gigantic gamble? Einstein very certainly knew the originators of the diverse concepts that he mistakenly interrelated into one idea; he revered these theoreticians as though they could not be wrong. Therefore, Einstein believed if he wove what he believed to be truthful propositions into one theory, it could not be wrong. Ah! A gamble, for critical propositions in the framework of his theory were wrong or fallacious and therefore general relativity was concocted on false premises. One very hard truth that was intentionally avoided is the fact that some of Einstein's concepts were only opinions and not based on any factual basis. Also the mathematics of his famous theory of general relativity is not a creation of his mind or given to him by a higher consciousness for the formula of James Clerk Maxwell (1831–1879) (and improved upon by Lorentz) was used by Einstein as a basis to weave the diverse concepts of others and by adding another coordinate to Maxwell's formula. Time is the coordinate added to James Clerk Maxwell's formula for electromagnetic radiation, and when it is peeled off, the Maxwell-Lorentz formula is laid bare.

In addition to this, there is another very important factor to consider in regards to Einstein's famous formula. Einstein's formula for the conversion of mass and energy is almost identical to the formula of Coriolis.

Did any of you readers know this fact? This fact certainly causes this researcher to be suspicious. I wonder; did Einstein have a powerful unconscious motivation to project his ego into scientific affairs in a manner to move into the limelight? As he interacted with the environment, his efforts revealed a desire to put his name on science and erase the name of others. Einstein's statement, "I stand on the shoulders of giants," in reference to where his ideas originated from is certainly true.

I have wondered, was this statement really a psychological ploy to use the work of others and put his name on it? The mathematics of general relativity was mostly James Clerk Maxwell-Lorentz. Only decades after Einstein as we say, died, was the deception discovered. The term *ether* was discarded by the manipulation of Einstein. Einstein said no one has ever proven the existence of the ether, "All our attempts to make the ether real has failed. Nothing remains for all the properties of the ether except for which it was invented; that is, its ability to transmit electromagnetic waves. After such unsatisfactory experiences, now is the time to completely forget the ether. It would be better if we never hear its name mentioned again."

The truth of this matter is outer seemingly empty space is actually a mass ether field although the amount of mass in hydrogen atoms is miniscule. Therefore, Einstein's reference to mass in empty space is correct in a small miniscule way, however, associating the occasional hydrogen atoms with energy is stretching a point, for energy is the capacity to perform work, and empty space cannot perform any work by itself.

Through Einstein's effort, his mass-energy-field became the new reference term in place of the ether. Einstein claimed the famed law known as the law of universal gravitation, developed by Isaac Newton, was not the force that held planets in place. Einstein claimed that a planet travels in a "space-time continuum," moving by inertia, following an "inherent curvature of space, taking the shortest possible route in a space that is warped and deformed."

At the time when Einstein was attempting to weave the theoretical concepts of others into his own theoretical hypothesis, two theories already existed as to the reason why the planets revolved around the Sun. Inertia was originally suggested by Jean Buridan. This theoretical scientist was a professor at the University of Paris. Buridan stands out in history for his stance against the Aristotelian concept of a necessary force to drive a moving body. Buridan postulated that a body once set on a moving course would continue on its course from the power of the initial impetus. He gave what he thought were examples to prove his conjecture, all incorrect. However, it was great theoretical conjecture since it was about 350 and some years before Newton. Newton incorporated the undeniable concept of inertia for stationary and moving bodies in his first law.

However, as the destiny of man unfolded, about 250 years after BURIDAN, the great Kepler postulated that a Sun force propelled the planets in their orbits. Kepler said, "There exists only one moving Soul in the center of all the orbits; that is the Sun, which drives the planet the more vigorously the closer the planet is" this force becomes weakened "when acting on the outer planets because of the long distance and the weakening of the force which it entails."

Ah! How I love that quote. Kepler and Newton were both probably aware of the relationship in square areas that is easily illustrated in plane geometry.

In concentric circles, where the radius is doubling with each successive circle, the area increases in ratio.

For example: if four concentric circles are drawn with a radius length that doubles in each successive circle i.e. a=2 ft. b=4 ft. c=8 ft. d=16 ft.

The area of a is r2 x pi or 12.5664 sq. ft.

The area of b is ~~~~~~~~~ 50.2656 sq. ft.

The area of c is ~~~~~~~~~~ 201.0624 sq. ft.

The area of d is ~~~~~~~~~~ 804.2496 sq. ft.

Therefore, as a is doubled and becomes b - the area between the two is 50.2556 / 12.5664 = 4, or four times greater. Thus, if an inch, foot or any increment of any substance (or in this case energy) were spread on an area, its thickness would have to be reduced by four in order to cover the b area. Understand, b would have ¼ (one quarter) the energy on the circle compared to a.

Similarly, c / a or 201.0624 / 12.5664 = 16 and c would have only 1/16 the energy spread on it compared to a.

Then d / a or 804.2496 / 12.5664 = 64. In this comparison, d would have only 1/64 the energy spread compared to a.

This also can be illustrated by squaring each radius to show how inversed square became the mathematical term. However, this time I illustrated the concentric circles, I was prompted to illustrate the reduction of energy away from the center point.

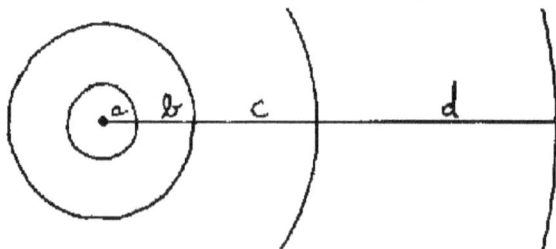

Therefore, as one educator stated, as you move three times a distance away and the attractive force, the force becomes one-third as much. You move four times the distance away and the force is one-fourth as much.

Newton was aware of these mathematical relationships and he knew this principle was not only true for plane geometry but was also true for spherical geometry. Since Newton was an astute mathematician, he used these concepts and Kepler's second law in assisting to develop the law of universal gravitation.

However, let's give credit where credit is due. Kepler was aware of these principles when he accurately claimed, "There exists only one moving Soul in the center of all the orbits; that is the Sun, which drives the planet the more vigorously the closer the planet is" this force becomes weakened "when acting on the outer planets because of the long distance and the weakening of the force which it entails."

Let us peer deeply into the above postulate of Kepler's. When the propositions of this work concerning energy and the planets of the solar system were given, the Aust and the three gravitational regions were also submitted. The regions were given as the vestal regions and this term was not carefully defined. We will now delve more deeply into Kepler's proposition as we define terms pertinent to the extension of his work.

Gravitational regions is submitted with this work as how it seems to me. Therefore, it is my speculation and is not submitted with the same assurance as the crux of this work: light energy dissolves into ethons, general relativity is a false theory, the nature of the electron, a rotating sphere creates an Aust or an extended rotary, gravitational field, an Aust enables orbs to be formed.

When mass turns or rotates as a gyro device or a rotating

sphere, it creates an extended force field within the region that its gravity field can control.

Vestal - This is an English translation of an ancient Latin term *vestalis* from Vesta, the goddess of fire. It is brought into modern usage by the guidance of an inner mind of my composite mind, to differentiate unique regions in the solar system.

1st or Inner Vestal Region - This term pertains to the physical solar system dynamics and does not have a meaning to the psychological influences emanated and generated by planetary movements. This first region is one of the three gravitational regions within the Aust created by the Sun. It contains the planet Mercury.

1st or Inner Vestal Region - This region is distinctly different than the subsequent bands because the first band of rotary gravitational force that girds the Sun has properties resembling physical substance.

Our solar system's center of gravity is located at the very center of the Sun and as the Sun rotates the rotary gravity force radiates from the Sun's physical composition. However, the extended rotary gravity field to the planet Mercury has characteristics of physical matter because of its intense gravity. At this region in all rotating Suns, the Aust may have the ability to bend light that passes through it from other suns.

As a wheel or a sphere turns or rotates, innumerable concentric circles surrounding the central axis have different rates of movement, since each circle travels through a different distance at the same time. A wheel can have the circles drawn or cut to calculate and illustrate this principle and a sphere can be cut into two hemispheres to mark or cut concentric circles for the same purpose. Therefore, the circumference of a wheel or sphere's surface region that is $90^0$ to its axis has the greatest rate of movement. However, with the Sun, extended concentric circles (or swirls since the Sun is rotating), extend away from the Sun to the planet Mercury as though physical substance occupies the plane polarized saucerized region 36 million miles away from the surface of the Sun.

The Sun's girth at its zero line of latitude is turning at the rate of 4,500 miles an hour. However, the rotary gravitational force traveling at 4,500 miles an hour is the basic driver that drives the entire revolving gravity field in the solar system.

This intense revolving gravity field of the Sun extends its rotary field force to the circumferential surface of the band at the inner limits of the 1st Vestal region to its outer limits and this is not revolving at 4,500 miles an hour but is revolving at about 107,082 miles an hour.

As distance from the planet Mercury increases, the subsequent bands move slower and are less intense, just as the swirling atmosphere moves slower and become less intense as distance increases away from the eye of a hurricane. Therefore, the outer limits of the 1st Vestal region band drives the next band and this drives the next until there is a coalescing where a tubular swirl or orbit is created. This band drives subsequent bands and this process continues until every planet is driven in the second gravitational region or the 2nd Vestal Region. Then the tubular vortex confine of Neptune drives the bands that coalesces or forms another orbit in the 3rd Vestal Region and this is Pluto. Therefore, each band moves successively slower as the distance from the Sun becomes greater.

This is why Kepler's law that $d^3 = p^2$ proves the accuracy of the Sun-planet relationship, and why the gravitational field become weaker.

Every moment the Sun is radiating gravity energy and

because the Sun is rotating, the radiated gravity energy has been forced into a saucerized shaped Aust where tubular swirls are orbits for planets. As gravity energy is radiated gravity, at the center of the Sun, Ethons of the ether are drawn into the central vortex and continue to replenish or to be radiated as gravity energy (pycnosis theory). That is very understandable and very logical, although, not every question about the nature of the orbital confines is answered.

A question that I have reviewed very thoughtfully is an aspect of the confines that involves whether the confines are permanent as atoms locked in a rock or is gravity energy that is radiated from the Sun continually replacing constituting energy concentration of the confine. Radiated gravity energy that has been formed into vortex concentrations could continually replace existing energy patterns in the Aust as those being replaced are radiated further away from the Sun in the Aust until the Aust loses its form. This may be a subject for experimentation in the laboratory using an artificial created Aust.

One key aspect is certain, and that is the gravity force becomes weaker as distance from the Sun increases. Thus, the planets move slower in each successive orbit away from Mercury. In round figures Venus moves 78 thousand miles per hour then in order, 66 - 53—29, 21 - 15 - and 12 thousand miles an hour for Neptune.

At this point the 3rd Vestal Region begins. Pluto has the distinction of being the only Planet in this region just as Mercury is the only planet in the 1st Vestal Region. These two planets also share the characteristic as having the greatest elongation, that is, their confine is the widest.

Pluto's great elongation causes it to move in front of Neptune and actually be closer to the Sun than Neptune when

Pluto moves to its perihelion. Due to this, I have often wondered if Pluto and Neptune could ever collide. However, they haven't collided in the last over 4 billion years, and with Pluto's 17-degree altitude, a collision is doubtful.

The sun's rotary Aust plexus, the Sun's extended rotary gravitational field or Aust has orbital tracks for revolving planets that are semi flattened doughnut shaped confines, (flattened torus). These confines for orbiting planets give birth to the astrological orbs. Each planet creates orbs in its orbit. This aspect of the solar system is not considered in this volume for it is explained in detail in Volume II, *Music of the Spheres*." From the astronomy perspective, there are three gravitational regions that I submit to be accepted or rejected by science.

These are: the 1st Vestal, the 2nd Vestal and the 3rd Vestal rotary motion in miles per hour for the Sun and each planet.

The rate of movement of our Sun at on surface at zero line of latitude, at the midsection bulge, is 4,500 mph.

Mercury's orbital speed is 107,082 mph end of the 1$^{st}$ VESTAL REGION

Venus' orbital speed is 78,337.44 mph

Earth's orbital speed is 66,638.16 mph

Mars' orbital speed is 53,977.32 mph

Jupiter's orbital speed is 29,214.36 mph

SATURN'S orbital speed is 21,586.32 mph

Uranus' orbital speed is 15,211.08 mph

Neptune's orbital speed is 12,146.4 mph

End of the Second Vestal Region.

Pluto is not calculated. This planet, as Mercury, is the only planet in its own Vestal Region. Pluto marks the end of the Third Vestal Region.

Newton relied on Kepler's second law as he analyzed

the Sun's force. His famous law of universal attraction with Kepler's concept of a weakening force with the increase of distance resulted in the law of inversed square. Although Einstein rejected the concept of Kepler, as did Albert Abraham Michelson (1852–1931) and Edward William Morley (1838–1923). Michelson's well-publicized perception of planetary movements that denied or ignored Kepler's concept of a central propelling force in the solar system was also a factor in Einstein reverting to the "inertia" concept of Buridan-Kepler.

While this is a fact, it is also recorded that Michelson never accepted Einstein's theory of general relativity. Why Einstein also denied the Kepler concept can only be attributed to Einstein's personal weakness in his ability to theoretically analyze in the realm of the abstract. To compensate for this weak ability he relied on the theoretical conjectures of other prominent theoreticians and tried to unite them with his idea of planetary troughs and his mathematical wizardry.

When a young man, a college professor had told Einstein to seek success in the domain of mathematics for in other realms he did not hold special strengths but in mathematics he excelled. Therefore, without a sharp perception of natural phenomena and the inability to supply a rational for his postulates, the die was set to create a gobbledegook theory.

When first trying to understand General relativity it seems to be very complex because Einstein knit together diverse and unrelated concepts.

The force responsible for driving planets around the Sun is certainly unrelated to the conservation of the quality in energy i.e. its ability to perform work, or is related to particle change when they pass through various forces, or causing matter to enter into space beyond the speed of light, without first dematerializing the matter.

However, Einstein's skill in mathematics as a maestro in music enthralled his peers as a violin virtuoso enchants his listeners. Einstein embraced the conservation principle of energy and postulated that mass and energy are indestructible and interchangeable.

He also postulated the amount of energy in matter is equivalent to the mass multiplied by the velocity of light squared or $E=mc^2$.

Please consider a very important historical fact. Thirty-six years before Einstein was born, a French scientist took up residence in a new environment, or as we say died, by the name of Gespard de Coriolis (1792–1843). This scientist is most remembered for being the first to recognize certain Earth forces to be altered by the Earth's rotation. While his work *did not* recognize the far-reaching gravitational affects from the rotating Earth. Of course every theoretician including Einstein missed or overlooked the vast effect produced from the rotating Earth, although, Coriolis did perceive that atmospheric currents were created by the rotation of the Earth. This condition is named in his honor as the Coriolis affect.

However, his conjecture that air movements above and below the equator being clockwise motions above the equator and counterclockwise motions below the equator were caused by the rotating Earth was only partially true. The clock and counterclock movements of air above and below the equator is caused by centrifugal force and gravity. In addition, high and low atmospheric pressures of air currents that pass each other, also causes clock and counter clock movements of atmosphere.

Although Coriolis was a brilliant theoretician and he also did other work concerning the amount of kinetic energy of a moving object and to find this unknown he submitted an undisputed formula. He claimed that the kinetic energy of a moving object is equal to one half the mass times its velocity squared or $E=m/2 \times V^2$. This formula certainly should seem familiar to all that read Coriolis' formula, for if a v for the velocity of the object instead of c for the constant velocity of light is used, the almost identical formulas are revealed.

$E=m \times v^2$ is known as Einstein's while Coriolis's law is $E=m/2 \times v^2$

$E = m \times c^2$ -- Einstein

$E = m/2 \times v^2$ -- Coriolis

It was bold to purloin Coriolis' law and bold to conjecture or speculate that $E= m \times c^2$.

I cannot know for sure that Einstein purloined Coriolis' law. However, that incident at Princeton, in the early 1950s, where Einstein was accused of purloining a fellow scientists' mathematical work about a unified field theory, and the similarity between Coriolis's law and Einstein's law, leads this researcher to suspect that Coriolis's law was the model for Einstein. In addition, the fact that Einstein's mathematics in his "Space Time Continuum" aspect of his theory of General relativity was developed by James Clerk Maxwell and Lorentz with a coordinate added, and that was time. Einstein never revealed this, and only years after Einstein died was the truth discovered. Therefore, I have good reasons to suspect that Einstein's famous law was a modification of Coriolis's Law. It certainly would be difficult if not impossible to prove that $E= mc^2$. for it could be substantially more or less.

Einstein "pasted in" or used in his theory of general relativity a theoretical conjecture originally speculated upon by an Irish and a German physicist by the names of George Francis FitzGerald (1851–1901) and the Dutch genius Lorentz, known as the "Lorentz-FitzGerald Contraction."

## TO ALL WORLD SCIENTISTS AND SCIENTIFIC ORGANIZATIONS:
## DO ALL OF YOU SOLEMNLY TESTIFY TO ALL THE PEOPLE OF THE WORLD THAT YOU ARE HONEST? YOUR STATUS AS HONEST DISPENSERS AND REVEALERS OF SCIENTIFIC FACTS IS AT STAKE.

Assuming that you claim to be honest, then answer a two-part question of the two following questions. A: Since Einstein's famous equation is almost identical to Coriolis's law—is it a coincidence or did Einstein use Coriolis's law with a slight modification? This question is important, for if Einstein used Coriolis's law then he purloined Coriolis's law. The second part of the question is B: is it just a coincidence that Einstein's famous equation is almost identical to Coriolis's law?

The second two-part question pertains to planetary motion. Since Einstein's theory of "general relativity" postulates that the planets revolve around the Sun by inertia and Kepler's third law was derived from factual data and states that the Sun is the driving force that propels the planets around the Sun in a precision movement so that each planet's distance cubed equals its period squared. A: was Einstein unaware of Kepler's law or B. did Einstein create a false idea—that was his *opinion* to justify or give false validity to his false opinion that stated "space is warped and deformed in the vicinity of matter" and the planets revolve around the Sun by inertia?

Also gentlemen: not only was Einstein completely wrong about the "electron emission affect" when light is focused on metal—for in this work—I submit - the emitted electrons were and are formed from the light wave that become absorbed into the atomic whirling's in the metal—they were not and are not emitted electrons from an atom's electrons. In addition gentlemen, tell the world that Einstein boldly and falsely stated, in 1932, the following: "There is not the slightest indication that [nuclear energy] will ever be obtainable. It would mean that the atom would have to be shattered at will." Both of these ideas of Einstein are great—hog wash.

Be honest gentlemen: for the first time in over a hundred years—I challenge you to be honest to the world's people and for end time of our history. Inform the peoples of the world that much of Einstein's false ideas were purloined false ideas from other scientists and he used these false ideas with his own false opinions.

**MORE ABOUT THE LIGHT WAVE AND LIGHT ENERGY FOLLOWS.**
The light wave has an interaction of its energy concentrations, (photons) within an alternating magnetic and electric field. This interaction is the motor that governs the rate of movement at 186,282 miles per second, independent of the light source. Even if light is passed through a gravitational or magnetic lens, light is deflected, or the course of light is altered, but the velocity remains constant, *if* the light is in the same conducting medium.

This understanding did not suddenly pervade scientific thoughts for testing and theoretical conjectures preceded the understanding that light has a constant velocity independent of its source.

Consider the following: if moving light could be quickened by adding the rate of motion of its carrier, we would see events happen before they occur or we could avoid events by riding a "time warp." For example: if two space crafts traveling at the speed of light are approaching each other on an angle towards point B where a collision is imminent, and the imaginary captain on one ship would order the imaginary crew to move to a launch vessels cabin, then fire the electromagnetic thrust to the velocity of light, since the space ship is already at maximum light speed, would the new light speed be added to the mother ship and propel the launch vessel at twice the speed of light therefore avoiding the collision. Very, very erroneous thinking for in normal circumstances light has a constant velocity independent of the light source.*

This illustration does not stipulate by preclusion the exceeding of light speed in a space ship by dematerializing it first, but we are a very long way from that condition. From another perspective, the news from CERN in November 2011 that a neutrino exceeded the speed of light should not be taken in a broad sense because a neutrino is not a self-propelled energy particle. An underground beam was driven by magnets in a 454-mile trip from Geneva to Italy and the artificially created force seemed to have driven the energy particles faster than light. I believe their measurements were flawed, I believe the speed of light was not exceeded.

**\*IN A FEW PARAGRAPHS YOU WILL READ WHERE LIGHT HAS BEEN SLOWED AND ACTUALLY STOPPED.\***

These illustrations could continue and occupy several paragraphs; however, the point here is the fact that this line of conjecture was coupled with conjectures coming from the Michelson-Morley experiments and more misunderstanding was generated. From this unintended confusion came the belief propounded by the Lorentz-FitzGerald baseless speculation that particles greatly accelerated would slow their movement, gain mass, that is, gain energy from some undefined and unexplained source, even though "energy cannot be created."

Dependent upon the perspective and the conditions associated with a perspective, a single event can have several explanations. For example: if a soft spongy slow moving electron that was created from red light was greatly accelerated, its "inner actions" would readjust to the new extreme conditions to preserve its identity or otherwise it would be blown apart. Take careful note scientific scholars, becoming denser is a condition that would preserve its identity. Either the electron would absorb energy from a magnetic field, or the accelerated electron would absorb—ether particles—ethons - into its own interactions.

You will find my own analytic scrutiny applied to the findings of this test and possible conditions, as eddy currents in a cyclotron that would alter an electron's mass and rate of movement. Take careful note: any eddy currents existing in a cyclotron would certainly mislead researchers to erroneous conclusions concerning the Lorentz-FitzGerald gain mass idea from which Einstein's "gain mass" postulate was copied from.

Follow closely, readers, for bit by bit, general relativity is sharply analyzed.

We now have an opening to expound upon the shroud of mystery concerning the time factor in general relativity that enabled science fiction writers to invent time warps; this concept is carefully explained so any ninth grade student will be able to understand the seeming baffling idea of uneven time.

Each reader should follow the expose' of Einstein's time concept carefully, for weaving this concept into his theory has confounded the minds of many and has resulted in one suicide of a physicist because he could not understand general relativity. We begin by defining time. Time is no more and no less, and simply a measurement of motion.

As soon as mankind began to develop scientific systems, a system of measurements was established. Therefore, to measure distance, weight, volume, power, and so on, standards of measurements were established. All measurements of time are equated to the Earth revolving around the Sun and the Earth rotating on its axis, and in Christian nations, the passage of years is measured from the accepted time of the birth of Christ.

Therefore, we count time, in years, from the year zero, measure time from the Earth revolving around the Sun as months, and finer increments as days, use the Earth rotating on its axis as hours, more finer increments as minutes, more finer increments as seconds, and tenths, hundredths, thousandths, and millionths of a second.

If we travel sixty miles an hour in our car, we travel a certain distance that is measured by one system of measurement, as the Earth rotated 15 degrees on its axis, since time is computed with the rotation of the Earth. If we travel for days, we calculate the time traveled by the Earth revolving around the Sun.

Time cannot be reversed or, to paraphrase, no one can go backward in time. If time could be reversed, the Earth would have to stop rotating in a counterclockwise direction and begin to rotate in a clockwise direction. In addition, the counter clockwise revolving motion would have to stop and the Earth would have to start revolving in a clockwise direction. If such an impossible event would occur, all life on Earth would be extinguished. However, if the Akashic records or the Book of Remembrance would be read, the records could be read as reversing a movie in a VCR.

For all that have science fiction ideas about time warps and folded space, understand that time is a measurement of motion, whereas one motion is correlated against another, i.e. a watch movement against the rotation of the Earth.

Before fuel injection and computers, timing an automobile engine was synchronizing motions that is, when spark plug firing is correlated to the piston being at the top of the cylinder. Before there were intelligent beings that were capable of making measurements of their physical environment, inches, meters, miles, kilos, Astronomical Units (AU), pounds and all units of measurements did not exist. Every system of measurements was created by mankind to conduct business, establish order and to control the physical environment.

Early man created the measurement of time as all other types of measurement to count the passage of days, lunar cycles and years. If man had never existed, no measurement system would have existed. Science fiction writers have had a heyday writing about fictionalized time warps, and time travel.

Magnifying glasses will enable one to enlarge or reduce the image one sees but by only recording events as with a camcorder or a motion picture camera can the past be viewed excepting Akashic records. The future can be calculated by using dependable repetitious events to foretell coming events, but they cannot be seen until they occur.

Our advanced hi-tech systems in our environment are dependent upon timing to very fine increments. However, when we turn our telescopes to vast outer space, time is out of sync, because it requires millions of years for the light of distant objects to travel to our Earth. So, we do not see real time reality when looking into outer space, we see *apparent* reality, Bruno stated this fact in his writings, yet we can make measurements by determining rates of movement and we use our created system of time to do this.

Before I continue, I wish to include some amazing information that I received via email February 2, 2,001 from a fellow scientific researcher. It is general knowledge that light travels, in round figures, 186,000 miles a second. When that fantastic speed is stated, two factors are given. First: the distance traveled is given (its motion) and secondly: the time required to move that 186,000 mile distance.

So here is an example of how motion and time are tied together or related. Nothing really astonishing so far has been given but now, pay attention. Several research scientists in Cambridge, Massachusetts, have caused light to slow to one mile an hour, then stop the forward movement of light and cause its time to stand still, that is, to be frozen or be completely arrested.

It has been known that the time factor in light's rate of movement is slowed when light passes through glass and some liquids. The researchers have taken this known principle to an extreme by actually stopping light and thereby *stopping its time* by creating a specially designed gas chamber.

Two researchers, working independently, Dr. Lene Vesterrgaard Hau of Harvard and Dr. Ronald Walsworth of the Harvard-Smithsonian Center for Astrophysics made this time-stand-still discovery. Two years ago, that was probably 1999; Dr. Hau was the first to slow light time to about 38 mph by passing it through a chamber of chilled sodium gas. One year later she was successful at slowing light's time to one mile an hour.

That success stimulated an urge to completely stop light and thereby completely stop its time and this was realized by another innovation. The special chamber of sodium atoms was magnetically trapped and chilled to a few millionths of a degree of absolute zero (-273 deg. C). Normally this condition would be opaque to light but by using a process called an electromagnetically induced transparency, that is, by shining a laser light through the gas called a coupling beam (a pathway for light to travel through) it enables a second laser pulse to pass through the chamber.

Now follow carefully, for the innovators sent pulses of laser light through the chamber and during the test the coupling beam was turned off while the probe pulse was passing through the chamber and the light stopped inside the chamber and did not exit the opposite side.

*The light had stopped moving and its time (measured by its motion) was stopped.*

Then the coupling beam was again turned on and the stopped light began moving and exited the opposite side. So, not only does time quicken and slow in matter when temperature is raised or lowered, time slows in electromagnetic energy when it passes through matter and can be caused to cease in controlled conditions.

Changing time to extremes in electromagnetic energy was not known in Einstein's time, but much was known about how time changes with changes in matter. Consider the following: in order to clarify the concept of time as a factor in general relativity, please consider the three states of matter plus the state of plasma. Any atomic identity can be used to illustrate the time factor, so iron, nickel and oxygen will suffice for illustrative purposes.

Iron and nickel are recognized as solids in our Earth environment, but each has a melting point where they will become liquid, a vaporization point where they will become a gas, and a plasma state (on a star) when the electrons revolve so fast and leap out to farther orbits until they escape the binding power of the nucleus.

Each state is dependent upon the thermal motion activity occurring at the particular state. Since time is a measurement of motion, time is greatly accelerated as thermal motion increases. Thus, as any atomic or molecular identity gains heat, the time within its atomic movement accelerates and conversely, when any atomic or molecular identity cools (thermal motion decreases) its atomic time slows.

In the case where oxygen (all gases) is continually cooled to a point where it becomes liquid then a solid, its atomic time greatly slows and at the point near absolute zero, its atomic time virtually stands still.

These simple facts concerning time in the atomic and molecular states of matter were woven into Einstein's general relativity to encompass all matter *if* it would be greatly accelerated in space. Einstein believed that atomic motion would slow (and therefore time would be slowed) if matter would travel through space with great acceleration and at the near velocity of light the mass would shrink in the direction of its movement the atomic motion would have slowed vary greatly and eventually stand still, just as a gas slows its atomic motion as it approaches absolute zero and as a solid its atomic time eventually stands still (becomes almost motionless.)

This is Einstein's concept that is taken from Lorentz and FitzGerald and is the coordinate added to the James Clerk Maxwell—Lorentz formula for electromagnetic radiation. Yet, this concept must be cross-examined by relating these postulated phenomena against a known true inertia law and that is, nothing changes unless acted upon by a force.

In all my readings about Einstein and his theory of general relativity I have not found any rationale to support his contention of mass increasing and time slowing as matter would be accelerated to the realm of light speeds excepting his dependence upon the theoretical concept known as the Lorentz-FitzGerald Contraction. Einstein believed that Lorentz could not be wrong and therefore did not cross-examine this idea as he didn't cross-examine the Buridan-Newton concept as the reason for planets revolving around the Sun.

It is a historical fact that Einstein tried to abolish the ether theory and replace it with a relativistic mass-energy field, but whether a body traveling through his theoretical relativistic mass-energy-field was supposed to have a bearing on "matter shrinking in the direction of its movement, gaining mass and time slowing" as it accelerated to the realm of light speeds, has not been found by my searches.

The Lorentz-FitzGerald contraction idea seems to have been sufficient to convince Einstein without cross examining its possible validity. So, as this researcher cross-examines general relativity and any possible force that would cause the Lorentz-FitzGerald Contraction affect must be considered. We know that a space ship must be shielded when it enters the atmosphere at rates of movement from about 25,000 to 17,000 miles an hour, because if not shielded from tremendous heat, the hull would melt and all combustible material within the craft would burn to ashes. So, friction from atmosphere heats the outer skin, but when shielded the occupants within are oblivious to the tremendous heat on the outer tiles.

So, if a space ship is propelled through space at the near velocity of light, it must pass through atoms of hydrogen and ether particles or for the moment—if you care to refer to space as a mass-energy-field, even though it is not an accurate reference to interstellar space, it must pass through the frictionless ether and strike few atoms of hydrogen as it moves.

This much is evident, but will the ether and hydrogen particles that are striking and flowing around the hull of the space ship have some type of power to cause the interior of the space ship to become affected in a manner to shrink the occupants (hey honey - I shrunk the kids) and slow their atomic and molecular constituents of their biological machine (their bodies) in a manner that would cause the aging process to cease or more realistically, cause instant death?

If you really believe this preposterous idea about slowing biological aging then you are among the adherents to the gobbledegook theory of general relativity and believe science fiction to be reality. As we continue to examine general relativity, please allow me to again consider the claim of Einstein pertaining to the uneven oval shaped orbit of Mercury. Einstein claimed the revolution of the ellipse of Mercury is a relativistic affect.

In my searches, I have been unable to find a record that Einstein ever magnified the eccentricity of Mercury's orbit to discover its quasi-egg shape. Therefore, it is not a wild conjecture to speculate that Einstein never magnified the eccentricity of Mercury's orbit (as you see in this work) to ascertain its true geometric shape.

Dr. John A. Wheeler, formerly of Princeton, proved that interplanetary gravitation (using Isaac Newton's law of universal attraction) is the cause for the revolution of Mercury's aphelion and perihelion points, therefore its uneven oval elongated orbit.

Also, as given before, a very important phenomenon pertinent to the concept of gravity that was completely overlooked or disregarded by Einstein is a phenomenon of clockwise and counter clockwise whirlpools of water (liquids) above and below the equator.

Even as I type near the year 2000, this phenomena is referred to as the Coriolis effect in honor of its discoverer, but the why has never been accurately explained. This isn't to say that scientists have not grappled with the Coriolis affect, and I submit to all readers a quotation from the second edition of an excellent and much respected *Encyclopedia Of Physics* by Rita G. Lerner / George L. Trigg, page 190.

"The apparent or "fictitious" acceleration to a body in motion with respect to a rotating coordinate system (e.g., the frame of reference fixed with respect to the Earth's surface) is known as the Coriolis acceleration. It is fictitious in the sense that if the same motion is described in a none rotating (inertial Newtonian) reference frame, no such acceleration is present. It is merely an apparent deviation from inertial motion due to the choice of a non inertial reference frame."

That particular encyclopedic explanation for Coriolis acceleration has several more paragraphs but the important factor here is the deft dealing with two systems—one that rotates and one that does not rotate.

For the scientific wizards: completely ignoring the counterclockwise and clockwise whirlpools above and below the equator evades the affect of a rotating Earth, upon the Earth's gravitational field.

For the readers in the northern latitudes that have watched the water in their kitchen sinks or bathroom bowls form a clockwise or a counterclockwise whirlpool as it is pulled into the lower plumbing system and have thought to themselves *it's the Coriolis effect*, they would be incorrect for gravity causes water to be pulled down a drain and gravity of a celestial body is influenced by the body's rotating movement.

Gravity's clockwise swirls north of the Earth's Aust, causes fluids to swirl in a clockwise direction toward the center of the Earth, unless acted upon by counter forces. The counterclockwise and clockwise gravity swirls are the strongest at the peripheries of the Aust. The gravity swirls are the basic or vital factors in causing hurricanes.

The width of the Earth's Aust should be mapped. Water does not swirl as it goes down a drain at the equator and does not swirl on both sides of the equator. Exactly where water begins to swirl as it goes down a drain north and south of the equator will mark the width of the Earth's Aust and the beginning of the gravitational eddy currents—that is, the Wake zone.

*Gravitational Eddy Currents Affect The Atmosphere And All Life.*

"Animals and birds instinctively avoid an area about 165 feet in diameter near Grants Pass, Oregon, called the Oregon Vortex where the laws of gravity don't apply. Inside the vortex, trees point to the magnetic north as compasses go haywire, pendulums hang at absurd angles, smoke blown to the vortex swirls as it is pulled to Earth and then vanishes as it strikes the ground. My 40-year-old notes pertaining to this phenomenon cannot be located, therefore, I haven't been able to find whether the swirling of smoke is counterclock or clockwise, at this "Gravity Vortex."

*Scoffers of the "Gravity Vortex" please read the next sentence carefully.* Contrary to many scoffers that claim the vortex is an illusion, when pictures are taken through the strongest gravitational anomaly, the developed pictures show distorted pictures. Remember how distorted a person appears at a fun house when looking at the convex and concave mirrors, also remember—on a hot summer day a distorted landscape can be seen when looking through a shimmering heat wave radiating from a highway. Also remember—light bends as it passes a star (special relativity) therefore—not only does light bend when it passes a star—light bends when looking through a heat field (as a hot highway) and light bends as it passes through a concentrated gravity field as the "Gravity Vortex" at Grant's Pass in Oregon. In addition to gravity's clockwise swirling movement in the northern latitudes as it pulls all particles toward the center of the Earth, there are other gravity vortices on the Earth.

Remember—Hurricanes Are Not Formed On The Zero Line Of Latitude.

Gravity is not just "a condition of space in the vicinity of matter," as Einstein claimed; gravity is a weak force in the electric-magnetic—electromagnetic—forces of matter. Each individual atom has a gravitational force that attracts other atoms and when two or more atoms join, they form a common center of gravity and attract other particles as though the two are one. In a larger aggregate of matter, the constituting particles join in a unified gravitational force and extend their attractive force to greater distances according to the total mass.

The gravitational force of a large body as a star, planet, or satellite, has a center of gravity at their central point of the body. Thus, the gravitational pull from the central point of all celestial bodies has caused all celestial bodies that were formed from globs of Sun matter to become spherically shaped, therefore gravity is not just, "a condition of space in the vicinity of matter."

Gravity is an inherent force of all atoms and the force

of gravity pervades the universe and it can be harnessed in a unique manner to make gravity engines able to propel space ships. However, from all the lessons of nature where energy is used to create a resultant force, two or more forces interact in a manner to create a resultant force. Therefore a gravity engine will have to have another force acting upon gravity to produce a resultant force.

**NATURE HAS MANY SECRETS, IT IS MAN'S FOLLY TO BEND NATURE'S LAWS TO FIT HIS IDEAS.**

Consider the following: when Einstein was working on developing a concept of the universe and most specifically, why Mercury and Venus were not pulled into the Sun billions of years ago, there were several well-known facts concerning the motion of celestial bodies that evidence reveals were not significant to his analysis. The shape of a spiral galaxy was known. Pictorial representations can be seen in old encyclopedias and astronomy books that were published before NASA was created.

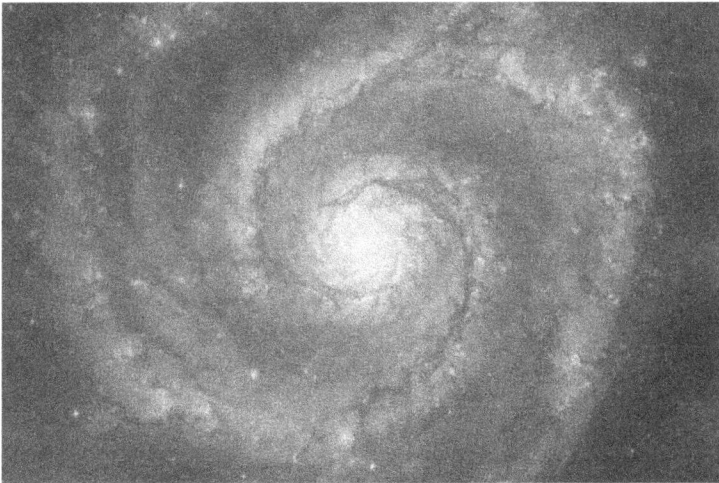

Globular Cluster below:

Also, the shape of a Globular Cluster was known to have a shape where its suns are clustered about the central Sun, in seeming random cluster arrangement.

If Einstein had seriously considered the true characteristics of these galaxies and their motions, he would not have used false assumptions that led to false conclusions. The following motions of celestial bodies were known:

1. the central sun of a globular cluster was not turning or, to paraphrase, the central sun of a GLOBULAR CLUSTER is not rotating.

2. the central sun of a spiral galaxy is rotating and the revolving radial arms that are spread out from the central girth of the central sun are revolving in the same direction as the direction of the rotating central sun.

Our own Sun rotates once in slightly more than 25 Earth days in a counterclockwise direction, and the planets also revolve in a counterclockwise direction.

(Astronomical data differs as to the Sun's rotational period, from 25.34 days to 25.7 days. It depends upon which latitude is measured).

Our Sun's planets also revolve in an area extending outward from the Sun's central girth.

Also, a rotary saucerized gravity field extends outward from all rotating spheres and this principle is the reason for the Earth's Moon revolving at the Earth's equatorial region in the same direction as the Earth's rotation. But this is not unique for a planet, for Jupiter's moons and the rings of Saturn can be given as other illustrations.

Jupiter has fourteen moons. The four brightest were discovered by Galileo and are named LO, EUROPA, GANYMEDE And CALLISATO. Jupiter's moons are divided into three groups. The innermost five revolve almost circular around Jupiter's equatorial region. The next four are farther from Jupiter and their orbits are inclined about 280 degrees to Jupiter's central girth. These nine moons all revolve around Jupiter in the same direction as Jupiter rotates._

The last four moons of Jupiter are very small bodies and revolve in opposite direction to Jupiter's rotation. This fact could be revealing something about a rotary gravity field or another possibility is their origin being attributed to captured meteors. If this be the case then inertia is responsible for their revolving motion, but if there is kind of reversal of gravity force, that could be the reason. If that be the case, then we could speculate that the reversing motion could be the cause for our Sun's Asteroid belt. Don't discard this idea for we more to learn about the gravity of a spinning celestial body.

Neptune has two moons, Its largest satellite, Triton, has a diameter of about 3,700 miles and is the third-largest moon in the solar system, largest being Jupiter's Ganymede and second largest Saturn's Titan.

Triton is the only satellite in the solar system that revolves in the opposite direction to the planets direction of rotation. Yet, it maintains the same positional orbit as all satellites since it revolves in an orbit that is inclined twenty degrees from Neptune's zero line of latitude. Why Triton revolves in a direction opposite to Neptune's rotational movement has yet to be discovered. Certainly at the present state of knowledge, a gyroscopic rotary gravity field has many secrets and in some future time, the reason why Triton revolves in a direction opposite to Neptune's rotation will be discovered.

Therefore, consider the facts about revolving bodies in our solar system:

1. not only do the spiral arms of a rotating galaxy revolve in the same direction as the rotation of the central Sun.

It is also true for the following:

2. all the planets of our Sun's solar system

3. the Earth's Moon

4. Jupiter's moons

5. Saturn's rings

6. Uranus' rings and moons

7. Neptune's moons

All the planets and moons named above revolve in orbits that are rotary gravity extensions from the controlling bodies zero line of latitude. All revolve in the same direction as the controlling bodies' direction of rotation, except Neptune's Triton. Some have theorized that Triton is a captured meteor. However, its shape reveals its primitive origin. If it's spherical, it was formed from a Sun glob when the solar system was formed. If it's a craggy irregular shaped body, it was formed from a collision or collisions of cooled Sun matter.

Neptune's moon, named Nereid, may also be a captured meteor. as well as the so-called moons of Mars. Their shape is the revealing fact, although this comes from my analysis, for I have never read where another scientist offered this scenario as to the origin of satellites.

These related motions between satellites and the planet of their domain plus the planets to the Sun adds to the concept of a uniform precision relationship between a dominant rotating body and its subordinate revolving bodies.

Trying to argue the possibility of happenstance or coincidence to attribute the binding relationship of each revolving planet and moon to its superior body is absurd. The speed of each of our Sun's planets is determined by its distance from the Sun, so that its distance cubed equals its period squared. Attributing this fixed relationship to coincidence or happenstance is utterly ridiculous.

Thus, from an additional perspective, the overwhelming evidence regarding the precision motions of dynamic celestial systems being tied together does not permit the idea of inertia within Einstein's general relativity to be considered as the force that is driving the rotating spiral arms in spiral galaxies or the force in our solar system that drives the planets around the Sun or the force that propels a planet's moons. Consequently, these truths reestablishes Kepler's concept of the rotary gravity field.

Shame on secular scientists that have the position to lead the world to higher development to be duped and "hoodwinked" by teaching general relativity to be actually real when it terribly distorts reality.

Not only did the above mentioned facts not have a vital significance to Einstein, these facts did not influence the minds of physicists and astronomers since the hype lauding general relativity given by the *Times* of London in 1919. Perhaps these scientists are of the same atheistic belief as Karl Marx. Many of these scientists became professors in schools of higher learning, and received their PhD by writing their thesis about general relativity.

In addition, there were men that had made important discoveries in 'electron emission' and had made erroneous conclusions in relation to these. Einstein and all the scientists working in organized science firmly believed the theoretical concepts credited to both Mayer and Helmholtz. Both of these German physicians turned physicists are given credit for leading scientists to believe their speculations that "energy cannot be destroyed." Ultimately, their beliefs was accepted as a law known as the "law of conservation of energy" i.e. "energy cannot be created or destroyed but can only be transformed."

Energy impacted in electricity became a factor in speculations after electricity began to be understood. Our history with electricity begins with the Greeks and their experiments with static electricity from causing atomic fields to interact with each other by rubbing amber.

In our modern time, many men contributed to our understanding of electricity. Michael Faraday probably has made the greatest advances in understanding electricity than any other man. But defining its nature is credited to the Dutch Genius Lorentz. In 1892, Lorentz suggested that electricity is the affect of charged particles flowing in a circuit and this correct speculation was meant for dynamic electricity—that is, flowing electricity, not static electricity. Yet, static electricity is caused by an excess of electrons that jump from a more highly charged substance to a lesser-charged substance.

Although naming the particle in an electrical current we know as the electron is credited to Sir Joseph John Thomson (1846–1940), even though an Irish scientist, George Johnstone Stoney, (1826–1911) had earlier suggested that a fixed minimum charged particle that he called the *electron*' was a particle of electricity, rather than thinking of electricity as a continuous fluid.

Although, it was Robert Andrews Millikan (1868–1953) who really proved conclusively that an electron is a particle, and he calculated the size of the charge of a single electron. (Droplets of water and oil experiment with the use of an ion).

Millikan's work does not detract from the importance of Faraday's work and it was Faraday that discovered the principle in the transformer of the "building and collapsing field and dissolving and forming electrons." While history records that Faraday did make the momentous discovery, he did not understand it, and the significance has lain in scientific limbo until this work has been presented, a period of about 150 years. I personally believe that Faraday received this knowledge from higher consciousness. Since I received knowledge from higher consciousness, my suspicion that Faraday also was a recipient of knowledge from God, Christ, or his Superconscious mind is reasonable.

Why did Faraday experiment with electricity in surges instead of electricity flowing as water in a garden hose can only be answered by the fact that only electricity in surges where its flow is interrupted by surges that resulted in a building and collapsing field in a transformer which enabled the current to be "stepped up" or "stepped down. In addition to myself, I suspect there were other metaphysicians in our history. The works of Faraday, Karl Gustav Jung, and Newton suggest to me that they were also metaphysicians.

Faraday's Work Served As Opening A Door To Greater Understanding.

However, Einstein did not perceive the significance of the principle in Faraday's discovery, but neither did Heinrich Hertz, Gustav Robert Kirchoff, nor the great German physicist Max Planck, on whose work Einstein also heavily relied.

If Max Planck had perceived the reality of Heinrich Hertz's experiment with a spark gap apparatus, i.e. the increase of electrons on a 'metal ball terminal' when violet light was focused on a terminal, our development would have taken a different direction.

Please read the following two reasons carefully. We have already given the explanation contained in the second reason. So, read the first reason very thoughtfully, for once you understand this principle of light being formed into electrons, you will have a greater understanding of this phenomena than did Max Planck.

1.  I submit - a small percentage of the violet light is absorbed by the metal balls and is formed into electrons by the internal whirling field motion of the atoms in the metal. This increases the number of electrons in the metallic balls atoms, or this excess quanta of electrons hastens the spark to jump across the air gap. Please reread number one and understand the forming of electrons from

light for this is a vital and a new concept presented in this work.

2. or if Max Planck would have understood the "dissolving and forming" electrons in a transformer, our development would have taken a different direction.

As it was, Max Planck's work with Kirchoff's theory of light absorption in a "black box" and Planck's concept of "black body radiation" was hampered by his misunderstanding concerning the nature of electrons, the dissolving of energy and therefore 'grievous erroneous conclusions' were reached.

Max Planck assumed (as Kirchoff assumed before) that all light energy absorbed by a black body stayed there as though being parked in a garage as a parked car until the black body was heated. At this time the parked light energies was assumed to again become energized and radiate from the black body. This concept is entirely incorrect. Max Planck did not have the least notion that light energy dissolves when a means is not provided for it to transform into another form; he firmly believed in J. R. Mayer's assertion that "energy cannot be destroyed" therefore he created an erroneous concept to explain the disappearance of light energy in a 'black box.'

In 1901, Max Planck's perspective of matter, light and electrical energy focused on the movement of particles whether it be particles postulated by Newton's 'corpuscular theory' or the electrons of Thomson. Planck knew that any energy absorption and radiation must be considered from the perspective of—'how much energy' and from this perspective he coined the term *quanta*, from Latin meaning "how much."

Max Planck was rewarded for his quantum perspective and the monumental work by being awarded the Nobel Prize in 1918. Max Planck's "how much" concept or quanta concept laid a firm ground for others to include this concept in their speculations. The forerunning work of Max Planck is an integral part of quantum mechanics as is the developers of the wave mechanics Paul Addison Maurice Dirac (1902–1984) and Erwin Schrodinger (1887–1961).

Quantum mechanics: semi-quantized electromagnetic and atomic forces. The conceptualized idea of an atomic force to hold atoms together by gluons is great theorizing, and from my analysis, is absurd.

I speculate that this idea will be dismissed in the future. The idea of bosons, quarks, pions and gluons makes great science fiction, but in truth, protons, neutrons and electrons that constitute atoms are held together by two atomic *cohesive forces* or *binding forces*. The first *binding force* is created by the electronic systemized particle interactions of the proton, neutron and electron.

The second *binding force* is the force that joins atomic particles to the Ether by a pycnotic action. When this dual *cohesive bond* or *binding force* is broken by fission or fusion, a tremendous amount of energy is released in heat and light, in an instant, known as an atomic explosion.

Einstein rejected the ether as a primordial neutrally charged electric field and pycnosis as an operating principle. In addition, as given many times in this volume, he firmly accepted the so called "law of conservation of energy."

With three basic principles concerning nuclear forces and actions changed to fit his concept of natural phenomenon, he incorporated his interchangeability of mass and energy formula that was taken from Gespard de Coriolis's formula for calculating the amount of kinetic energy of a moving object and submitted a formula that was supposed to be the amount of energy in million electron volts that was inherent to atomic *cohesive force.* This was pure speculative conjecture just as his idea that energy cannot be dissolved and the planets revolve around the Sun by inertia.

Before we wander too far, let us return to Max Planck's great contribution—that is, his correlation of mathematics to actual phenomena. He submitted a generalized formula for electromagnetic radiation that revealed the quanta of energy in light energy is directly dependent upon its oscillations or frequency. This generalized formula is true for any given light intensity. Thus, the quanta of energy is inversely proportional to the frequency of the wave. The violet end of the spectrum being the shortest wave length, having about twice the number of oscillations as the red end, and about twice as much energy; therefore, an inverse constant exists.

Planck's quantum theory did advance science a giant step. His specific focus where a generalized mathematical relationship between light frequency and the amount of energy inherent or within that specific region of the light wave, that is being inverse, was correct. It was his general perception that did not reveal to him the truth of reality, i.e. energy dissolves and the nature of the electron as illustrated in this work, as again given in the two following illustrations.

1.  In order to transform, that is to inversely alter, EMF and the number of electrons (amperage) in an electrical circuit, a building and collapsing field in a transformer dissolves and reformulates electrons and EMF.

2.  Plus, Heinrich Hertz's experiment with a spark gap apparatus, i.e. the increase of electrons on a 'metal ball terminal' when violet light was focused on a terminal, decreases the time needed for a spark to jump from one terminal to the other.

Even though Max Planck never knew that light could be formed into electrons, he was deeply interested in the amount of work that could be done by harnessing the energy of the light wave. This research led him to bring forth two formulas that pertain to the amount of work that can be done by the energy in the light wave. The one formula that incorporates light intensity may be correct and I am not in a laboratory to personally check this formula. But the first formula that was devised by Planck was dependent upon a constant which became known as Planck's constant.

The constant is represented in the formula by a small—$h$—and this formula is supposed to state how much work can be performed when the frequency of the light is known and inserted into the formula. This formula is stated as, $h=6.626\text{x}10\text{-}34$ joule-second. My great yearning has always been to prove or disprove a postulate or formula and in the case of Planck's constant, I was never able to make direct measurements with various intensities of light to prove or disprove its claim. I am a theoretical physicist and I have done my work from my office and library at 624 W. Princess St. in York, Pennsylvania, and 2270 Herman Drive Shiloh Pennsylvania. However, if I could make direct measurements with sensitive instruments and discover that Planck's constant is accurate, great inaccuracies still exists in his ideas about light in a black box.

So, Planck's general understanding of light and energy is erroneous.

1.  he believed that light can be parked in the surface area of a "black box"

2.  he did not understand that "light dissolves" and therefore when "light dissolves" the energy in light ceases to exists

3. he did not understand that light can be formed into electrons.

Many researchers working with cathode rays, magnetism, electromagnetism, electron emission, an electron flow. Cosmic rays and cosmic showers contributed heavily to the advancement of physics and the development of the electrical and electronic sciences. Einstein tried to understand the photoelectric effect, (without success) that is, the electron emissive affect when light is focused on certain metals as did a contemporary by the name of Lenard (lay'nahrt) (1862–1947).

In 1902, a year after Max Planck's introduction of the quantum aspect, which broadened the concept as to how much energy is within a given force, another scientist (mentioned previously) by the name of Lenard began to study the photo electric affect that dates back to Heinrich Hertz.

Lenard's work was focused on the 'emission of electrons' by metals when light contacts the metallic surface. Lenard found a relationship between the 'emission of electrons' from the perspective that as light varies in intensity so also varies the numbers of electrons emitted. However, he found that all the electrons possessed the same energetic state.

Einstein was also interested in the photo electric affect and it is this work though being correct from the quantum perspective of Max Planck's (the amount of energy in the light wave is inversely proportional to the frequency of the wave) *nobody*, including Einstein, understood that light is transformed into electrons. Einstein was *incorrect* as to why electrons were emitted by certain metals when light was focused on the metal. Einstein's error was due to his not understanding the phenomena.

*Electron Emissive Materials*

Einstein claimed that a particular wave length of light (being fixed in energy content or quanta of energy) would force out an electron and electrons from the atomic structure of the emissive metal that is equal to the particular wave length contacting the metal. This concept is completely erroneous; Einstein was 100 percent wrong.

Einstein speculated that the brighter the light the greater number of electrons would be emitted, this is certainly true, however, he also speculated the following: as the different colors of light or wave lengths of light are used, the 'forced out' electrons from the atomic structure of the atoms would be in accordance with Max Planck's formula. This is completely false for in this work I prove, beyond any doubt, the emitted electrons were and are formed from the electromagnetic light wave. Therefore, Einstein's conclusions were false, yet it gained him a Nobel Prize in physics in 1921. If the emitted electrons did come from the atoms of the metal, the metal would have become electrostatically charged and this has never been detected and I state - it never happened.

So, Einstein's first false claim, even though his false claim, which gained him a Nobel Prize, was:

1.   He claimed that emitted electrons came from the electrons of the atoms of the emitting material.

2.   The second absolutely ridiculous false claims were incorporated in his theory of general relativity.

It was during this work with electron emission that Einstein verbally speculated as to the nature of the electron when he said, "Perhaps the electron has an inner action." Absolutely the electron does have an inner action. Remember the explanation of the building and collapsing field within a transformer that dissolves and reformulates electrons? My research and analysis of the operation within a transformer by 1960 had revealed to me that 'non atomic' electrons are formed from a collapsing and building field motion of an AC current within a transformer. The process is so easy to understand. The breaking of electromagnetic waves into individual phases, as in the photoelectric affect is performed by the whirling field motion in the atoms. This creates a closed circuit from each or two individual phases and forms a particle or forms an electron. Thus - the dynamics that propel the electromagnetic wave forward is caused to produce a spinning electron.

If violet light be focused upon a photoelectric crystals or a photo emissive metal, the electrons formed would be hard and energetic having a greater quanta of energy. If the light be at the lower end of the spectrum the electrons formed would be softer and less energetic having lesser quanta of energy. The full light spectrum or clear light has the seven steps in its wave structure which can be individually seen through a prism or a rainbow. However, the seven different vibratory steps are also inherent to the light wave in red color, blue or any color. In these separate colors the wave structure is confined to a narrow range.

Electrons that flow in a circuit are equal in energy content, dependent upon variable factors at their point of generation. Free electrons have a range of energy concentration and after absorbing sufficient energy to move out of the electron class, the stable heavy particles are referred to as mesons and leptons. Leptons are not stable and break down into sub atomic particles. Tau lepton, for a stable heavy electron has been added. Naming a heavy electron a tau lepton earned the American physicist,

Martin Lewis Perl, a Nobel Prize in 1995. He shared the prize with Frederick Reines, the discoverer of the neutrino in 1950.

Before 1960, I had understood the original structural form of electrons to be a closed circuit of a phase of electromagnetic waves, also I firmly believed that in a nano second the formed electrons probably assume a spherical shape.

Years after 1960, I read that August Kekule Von Stradonitz (1829–1896) had also solved his problem concerning the nature of the benzene atom by speculating upon his problem while he was dozing and dreamed of a closed circuit or the atoms forming a continuous circle.

According to Isaac Asimov in his *Biographical Encyclopedia of Science & Technology*, Kekule saw atoms whirling in somewhat of a dance, when the tail end of one chain connected itself to the head end and created a spinning ring.

Another historical article that I read stated that Kekule perceived the spinning atoms to be as a snake with its tail in its mouth, thereby forming a circle. Through Kekule's work we are given the method of writing structural diagrams for chemical formulas. This innovation of Kekule enabled chemistry to nearly explode with expansion.

In regard to the photo-emissive effect, one great truth that escaped Einstein and has escaped or eluded the accepted scientific world for over 150 years is the fact that an electrostatic charge would be created when millions, billions and trillions of electrons would be 'forced out' of the atoms in matter; this condition has never been detected in a 'photo-emissive' metal or electrical transmission lines. Because electrons *are not* forced out of atoms even though Einstein believed electron were forced out of atoms and thousands of physicists have believed and do believe this false idea.

In 1958, '59, and 1960 when I was focusing on the nature

of electricity, I perceived that hard and soft electrons that are moving fast and moving slowly from the photoelectric affect are formed from the inner workings of the different frequencies (colors) of the light wave and when the electrons in the whirling atomic structures become, in excess, an emission occurs.

Electrons formed at an electrical generating plant are formed from a building and collapsing force field as field in matter cuts across the lines of force in a generator and the electrons and the EMF formed being motive take the line of least resistance and follow the transmission medium (lines) that have been provided for it. Thus, electrons flow through matter in a moving electromotive force field, and that is electricity.

We might be led to believe that we are finally closing in on the nature of the electron by now having a perspective of its nature expanded. Within the consideration of this, we may be led to think: the electronic industry including the advancement of computers has been made possible by the belief that electricity is a flow of electrons and now with a greater understanding of electricity doesn't this close the door on a chapter of science? I do suspect the ELECTRON has yet another aspect to its nature that we have never suspected in our wildest dreams.

I submit in theoretical conjecture that an unknown accompanying wave structure inherent to the entire electromagnetic spectrum and in all energy particles that enable the phenomena of dematerialization and materialization to occur.

Let me make this very clear to you. My work does not hinge on my suspicions concerning an aspect of reality that may seem to many as science fiction. We must be mentally sharp about reality and not my speculations. Therefore, consider the following: electrons are formed by our contrivances by five methods. They are as follows:

1. A generator or dynamo

2. photoelectric cells

3. electrons formed from heat

4. pressure on crystals

5. chemical devices also electrons are captured by devices from 'radioactive decay.'

1. Electrons are formed by a generator when the field in matter is caused to cut across electromagnetic lines of force, thus creating a continuous collapsing and building field and consequently forming generations of electrons.

2. Electrons are formed by the photoelectric effect by three types of devices. They are as follows:
    A. photo emissive or photo electric—illustrations have been given to illustrate how electrons are formed from the light wave.
    B. photo voltaic.
    C. photo-conductive.
    The photo emissive or photo electric principle is traced back to Einstein and Lenard in their efforts to understand the photo emissive metal when light is directed on the metal, also this principle is related to electronic tubes. When light impinges on a metal plate that has the photo emissive characteristic, it causes the cathode to emit electrons to the anode in the tube. It is a weak flow due to the small photo emissive surface area in the tube. Stronger current have been realized by a successive relay system that forms additional electrons from additional electrical charged plates.
    These additional plates or dynodes as they are called, adds

a secondary surge of electrons with the quickly relayed beam of the original electrons. Because of the second step relay system used to multiply the number of electrons, the multiplier process is called—secondary electron emission.

B. Photo voltaic devices are photo electric cells that convert light energy directly into a current of electricity, i.e. electromotive force and electrons.

C. The photo conductive cells are similar to the second type since both will convert light directly into a current flow.

The photo conductive cell is different from the perspective that it is a semi conductor. The photo electric affect depends upon a diminishing resistance in the semi conductor that is in proportion to the amount or intensity of light falling on it. As light intensity increases the resistance in the photo semi conductors decreases and electrons flow through the path provided for the electron flow.

A. the number of electrons emitted per second from a photo sensitive surface is proportional to the strength and color of the incident light

B. the maximum kinetic energy of each emitted electron is dependent upon the light intensity, (the emitted electrons will be few or many) and is directly proportional to the light frequency (will be hard or soft.)

3. Electrons are formed from heat. Thermocouples are used in an electrical circuit to convert heat energy directly into electrons and thereby a flow of electrons. Thermionic emission is a principle responsible for 'electron emission' in an electron tube (the idea that electrons are "boiled off" from the atoms when the plate gets hot is a false idea).

When a flow of current through the cathode that has been

coated with an emissive material, raises the thermal activity to a point where the electromagnetic radiation is close to the lower end of the light spectrum, electrons are formed by the whirling field motion within the atoms.

4. Electrons are formed from pressure upon crystals. Rather than explain this phenomena within the framework of my work that is presented here, I offer a theoretical conjecture. Piezoelectricity may come from the forcing of a condition where a collapsing and building field occurs.

5. Electrons from chemical reactions that is batteries and fuel cells do take electrons from atomic structures. This is different than the aforementioned methods of producing an electron flow as is the following.

6. Electrons can be captured or directed to flow in a current by using a device to direct ejected electrons from radioactive decay (beta waves) into a conducting medium. Electron flow and the forming or origin of electrons has been explained in a simple manner. Yet, with such clarity that present held beliefs including the theory of general relativity should easily be understood to be false.

## As a very young boy I wondered about electricity and lightning.

When I was five and six years old I wondered where lightning came from and how electricity could travel through wires. I am now 66 years old (it is now December 1994) {as I edit my manuscript as this time—I am 86 years old and it is now June 2014} and I have known the answers to these questions for decades. Science has classified lightning into types and forms. The forms are: sheet lightning, ball lightning, St. Elmo's fire, glow discharge into the surrounding atmosphere and bolts from the blue. The bolts from the blue are the visual jagged, forked or zigzag forms of electricity that can be compared to the arc seen with a static electricity machine. With this laboratory device, an arc of electricity can be seen to jump from one ball or terminal to another.

In nature, the bolts from the blue can jump 1. from the front of the storm to the back of the storm; 2. from upper clouds to lower clouds; 3. from a low rain cloud to the Earth; 4. from a squall cloud (severe local swirling rain clouds) to the Earth, and 5. from an upper cloud to the Earth. These understandings were derived by direct observation. However, a definitive consensus as to the origin, generation, vehicle, or medium that generates the electricity has never been definitely agreed upon by physicists or meteorological scientists. The reason for this lack of agreement among scientists is also related to the science of physics, and here again the acceptance of the theory of general relativity has interfered with the understanding of natural phenomena.

Consider the following explanation. Clouds contain moisture; when this moisture condenses, it produces rain. If two clouds of differing barometric pressure pass each other, it causes a huge eddy current in the sky. When a mass of oxygen atoms swirl in a circle it causes the atoms to cut across the Earth's magnetic field. This causes a building and collapsing of magnetic field and a creation of new electrons. When the atomic structure of oxygen continually cuts across the Earth's magnetic field a 'building and collapsing' of the field causes excess electrons to be formed.

As electrons collect in the cloud an 'electrical pressure' is created because electrons are negatively charged and electrons are motive. Electrical pressure can only build to a certain extent, that is, the capacity for the cloud of gas to hold excess electrons has limits. When this saturation point is reached, a charge of electrons will jump from a more highly charged position to a lesser or weaker charged position and that is a lightning bolt.

Lightning Is Formed By A Revolving Armature In The Sky.

Did you wonder how oxygen can act as an armature in the sky? Did you know that liquid oxygen can be induced to hold a state of magnetism? It can. When I was about twelve years old, I put aside building model airplanes for awhile and began building crystal sets to listen to the radio from a set that I had built. The unresolved question about how does electricity travel through wires came to the fore of my cannot get rid of these lines thinking whenever I became in contact with a knowledgeable person in the radio business.

At that time the best explanation that I also thought may be true was the fact that the best conductors of electricity all have a loose electron on their outer shell. However, was it true after one considers the fact that water also conducts electricity? I knew that the molecule of water was a stable molecule unlikely to give up electrons. About 18 or 19 years later, when I became absorbed in the study of the solar system, I analyzed the truth from many facts that were available from text books and finally understood how electricity travels through wires.

Electrons must travel through and with electromotive force. Electromagnetic radiation in space travels by a propagation of expanding spherical shells. However, electromotive force in an electrical circuit is confined to a conducting medium, usually wires. So, in AC and DC current, electromagnetic radiation travels with an alternating force field within the wave system, when the magnetic field is in, the electric field is out. Therefore, the one builds, reaches a peak then diminishes and goes out, *but* as the one decreases the other is increasing and it builds to a peak then diminishes until it goes out. This alternating cycle repeats itself continually as the electromagnetic force generates.

It is a similar building and collapsing field that causes electrons to be formed and dissolved in an AC transformer, in a generator and alternator as the armature cuts across magnetic field, and in the sky as swirling atmosphere cuts across the Earth's magnetic field.

Direct current cannot operate in a transformer because the needed cycles of shutting off and on the flow of energy that cause the building and collapsing field does not exist. Direct current is a steady flow as water flowing through a hose. Therefore, without the necessary building and collapsing field action, DC current cannot be transformed either in a 1:1 relationship or any ratio change.

In a step down and a step up transformer of an AC circuit, electrons are formed (they increase in numbers) as EMF or Electro Motive Force decreases and conversely as electrons are dissolved, EMF is increased.

So, you may ask, what does this have to do with how electrons travel through wires? Electricity travels through a medium that does not have a loose electron on the outer shell of the constituting atoms, although, it encounters greater resistance in its flow.

Thus, a loose electron on the outer shell is not a prerequisite for a flow of current. A loose electron does facilitate a collapsing and building field motion. Since electricity does not flow in a transmission medium without EMF, we must conclude that electromotive force is necessary for electrons to flow in a transmission medium.

So scholars and seekers of truth, burn deeply in your memory the following: Due to the great differing characteristics in matter, atoms with a loose electron on the outer shell have an inherent 'excellent capability' to conduct Electromotive Force. This is true for normal temperatures found in every clime on Earth. However, at near absolute zero or Cryogenic temperatures in some materials, superconductivity is achieved.

In logical extension, it is a created situation that prevents atoms from free movement, thereby creating an improved conducting medium. Another factor bearing influence to this line of reasoning, is the fact that super, supermagnets use samarium atoms to "fix" alignment of magnetic atoms to create an unobstructed flow of magnetic force out of its north pole and into its south pole. This is mentioned for all students to understand the comparative principle operating that enables electricity to flow almost unrestricted.

Yet, there is a principle beyond this and great research is being conducted that will achieve superconductivity at room temperature in materials using the fixed crystalline arrangement in certain substances. The goal is to use the natural space lattice formation of crystals to fix a permanent conducting field that offers a super excellent capability to conduct EMF and electrons on par with cryogenic methods.

So, the *idea* that electrons travel in wires as though a relay or 'train of electrons' is created where electrons start to travel and push out 'outer electrons' as the next behind takes its place and a great continuous chain of electrons move along the wires, is an old untrue idea. In a direct current, electrons travel with a flowing electromotive force; in an ac current, electricity travels in continuous surges and electrons travel along with and in electromotive force. Also, as the cycles increase, the flow of current moves farther outward towards the wires outer surface i.e. The skin effect.

Therefore, there are two conditions that create the ideal environment for electrons to travel in a conducting medium. The first condition is in direct current where a flow of electrons flow as water through a hose with electromotive force (voltage) and do encounter resistance in a conducting medium.

In an ac current, the outer loose electrons of the conducting medium facilitate a collapsing and building field of the electromotive force (EMF), *if the current is sent through a transformer.* A better name for an alternator is a surganator. The reason is obvious, an alternator creates continuous surges of electricity.

The second condition occurs with alternating current where the current flows with the EMF in surges and can reach super conductivity' where resistance disappears. In this condition electrons travel in the flow of electromagnetic force and the building and collapsing of electromagnetic field is not hindered.

In review, there are three points to remember concerning alternating current.

1.   In alternating current circuits the very center of a wire carrying an electrical current can be compared to the 'eye of a hurricane'.

2.   Electrons and EMF or voltage travel on the surface of wires in an AC circuit and this condition is referred to as the skin affect. Also, the higher the frequency (the greater number of phases) the greater the concentric skin effect expands, requiring heavier insulation.

3.  This does not challenge Ohm's Law, where it is stated: the resistance is directional proportional to the cross section, for Ohm worked with direct current and his law tied I, E and R, together i.e. (amperage, voltage, and resistance).

In an AC circuit, the resistance of a circuit is called reactance. It is the opposition to the flow of AC that is caused by the capacitance and inductance in a circuit. You can think of inductance as similar to resistance or the ability of a conducting medium to permit the flow of electromotive force. Capacitance or storage ability occurs in an AC circuit from the building and collapsing field of the electromagnetic force and creates an opposition to the flow of electromotive force. These together constitute the total resistance or reactance in an AC circuit.

Ohm's work was not honored in his native Germany, but English scientists saw the importance of his work, and in 1841 Britain's prestigious Royal Society awarded him a prestigious medal, and in 1842 they made him a member of the Society. This gained for him scientific acclaim, even while his German colleagues had cast him out of his teaching position.

Ohm was finally honored by his German colleagues and was given a teaching position in 1849 at the University of Munich by Ludwig I of Bavaria. Ohm's laws led to practical applications for electrical energy, and were truly the necessary mathematics for the founding and expansion of electronic science.

I trust that my presentation as to the nature of electricity and how electrons travel in wires will answer many questions. However, the absolute nature of the electron may not yet be understood for inquiring questions concerning its absolute nature still raises other theoretical questions.

The physical composite nature of electrons has previously

been revealed by their dissolving and forming in a transformer. There is more to understand concerning the full or complete nature of electrons. Consider the following question: Do electrons travel as a piece of wood flowing along a fast moving stream or do electrons dissolve and form as they move along with and in the EMF? The answer to that question follows.

According to my concept and the pycnosis theory, electrons are absorbing ethons from the ether and radiating energy as they move in various environments.

Within the given concept of electrons, the pycnosis theory and the Newton-Cavendish gravity field, the following theoretical question can be considered.

Can electrons be formed from a gravity field? To discover the answer to that question, an experiment can be performed by setting up the experiment within ideal conditions. These conditions exist at Grant's Pass in Oregon where a Gravity vortex exists. If electrons can be formed or to use a paraphrase, if electricity can be generated from a Gravity Vortex not only will my understanding of electrons be proven but the true nature of gravity will be better understood and the idea of gravity being a relativistic effect will be positively proven to be a false concept and thrown out forever.

The method for this experiment follows. By using the swirling vortex gravity force as a revolving stator and building a special stationary rotor that utilizes the swirling gravity force to cut across its windings—electricity should flow, if gravity is a type of electromagnetic force. If the results are positive from the test explained, the science of mankind will indeed have taken a giant step forward and another reason why Jack Truett Sr's work will be remembered forever.

Two other conditions of nature that were generally known at

the time Einstein was developing his concept of reality concern heat and gravity. Down through the centuries and up to the time when Einstein grappled with aspects of reality and up to the present time, it is known how the air shimmies and weaves over a fire. When one looks through the air that shimmies over a fire, one sees their surroundings in distorted shapes. That situation is only possible by the bending of light. So, if a thermal field can bend light, can other force fields also bend light?

From the work of the great Lorentz, with the help of another Dutch physicist, Zeeman, we know that a magnetic field can affect light, and it is said in 1919 when photographs were taken of the starry heavens a displacement of a star was found that led researchers to believe that 'gravity bent light.' However, according to my understanding, the picture was focused on the north pole region of a star and this is where the magnetic field is the strongest.

Certainly gravity of a black hole can bend light but for ordinary stars it is just as reasonable to attribute magnetism as the source for bending light in addition to gravity. Although, as I stated previously, the Aust may be able to bend light and a test in a laboratory where an Aust can be created, and will add empirical knowledge about the properties of matter and forces. This should clarify a situation now in question as to whether it is gravity or a magnetic field that bends light as it passes a star. I state that both forces can bend light.

We do know our Sun's magnetic field is about five times the strength as our Earth's magnetic field and the strength of each star's magnetic field is different. Einstein knew that celestial bodies that are not self luminous are not shrinking, so in his reasoning, if gravity would be a force in the family of electromagnetic forces, it would seem that great quantas of energy should be radiated away from the body resulting in a loss of mass and this has never been detected and it never will be detected because of the pycnotic action.

Secondly and more complex, Einstein was confounded the same as all theoreticians concerning a radiated force that would pull instead of push. To solve this theoretical dilemma, Einstein conceived an inherent power that permeates all creation to be the source of the pulling power where matter exists. Indeed he was knocking on the door concerning the creative energy of the universe whose particles are ethons.

However, the creative substance of God and the same energy that constitutes the human mind (all minds) was beyond Einstein's ability to fathom, for this concerns the nature of God. Nor could Einstein comprehend the expansion of consciousness that occurs when the Father Within becomes unified with the mortal mind as occurred by means of Jesus the Christ.

Einstein's theoretical conjectures never included positive comments in regard to the Infinite Mind of the universe nor did he submit any writings concerning the operation of the Infinite Mind. He simply gave up trying to understand, for he closed his mind in these matters and did not accept the presence of God (an infinite mind) or God's creative forces.

Since Einstein's theoretical work in physics was largely formulated by conjectures of other theoretical thinkers and the idea of a gravity wave having a screw affect within its wave structure that would pull upon other gravity waves that it contacted, never became a strong possibility in theoretical reasoning. However, this is the manner in which gravity waves interact with each other, and this screw effect is the root reason that governs the speed of a free falling object which is known as terminal velocity.

The terminal velocity is different on every celestial body,

depending upon its mass, on earth the terminal velocity is 250 miles an hour. So, if Einstein were told to disregard the ether and disregard the pycnotic concept of matter and develop a theory that includes gravity, J. R. Mayer's theoretical concept, the fact that light bends, build it upon James Clerk Maxwell's-Lorentz's mathematical law for electromagnetic radiation, use the inertia concept of Buridan-Newton that causes the movement of heavenly bodies, and use Coriolis's formula for the amount of energy in mass excepting don't divide the mass in half, include Max Planck's black body radiation concept with the FitzGerald-Lorentz gain mass shrink in the direction of movement false idea and use Giordano Bruno's relativity principle, those conditions taken from other theoreticians, would have created diverse theoretical concepts unsuited to be woven into one theory.

The use of Bruno's relativity principle hyped Einstein's theory for only a few scholars remembered the works of the Italian roaming philosopher that was burned at the stake.

These guidelines were used in his general concept that became Albert Einstein's theory of general relativity.

It is the greatest maze of gobbledygook ever devised by the mind of man, and the greatest amount of purloined material linked in a fantasy than ever before with Einstein's name affixed. No invention or device that would enable scientists to change or harness the forces of our environment ever came from general relativity and never will because general relativity is a false theory.

Let us now use our cunning to gain greater control over our environment and advance humanity. Let us now move into the area created by a new and truthful perspective. Gravity should be able to be examined with a new and sharp focus. If, electrons can be directly formed from gravity, it will be a giant step towards controlling and capitalizing upon our environment in a beneficial manner by harnessing gravity to create gravity engines.

If this type of theoretical reasoning seems too deep in the abstract, pause for a moment and consider the following.

At this time 1995, gravity engines, brain-cooking beams, Groom Lake, Roswell and Socorro, New Mexico are on the mind of many theoretical scientists. However, as you search to discover what is known and what is concealed, do not be deceived by the TV motion picture about the autopsy of space aliens in New Mexico because it has no basis to relate it to an actual autopsy of space aliens, although it may be a clever reproduction of an actual autopsy of space aliens.

After I contracted osteomyelitis in my right tibia when I was five years old, my physical activities were greatly reduced and I became a questioning little boy. As I grew, my life was increasingly driven by a profound curiosity. After getting married to my lifelong soulmate when she was seventeen and I was eighteen, I spent my life trying to be a good husband and father and always tried to go the extra mile for my employers.

I was never in a fist fight, never sat in a bar and drank, never saw a pornographic magazine or a pornographic movie. I spent my life studying and thinking in my spare time. I never went fishing, hunting, or went bowling or to a baseball, football, soccer or any type of game excepting to a couple basketball games while I was in school. I have a strong empathy to the culture of the plain people. Among those are Mennonites, Brethrens, and other plain people.

I was usually scorned for my strong work ethic, and when coworkers were successful at forcing me to engage in a conversation, I was ridiculed for my intense interest in metaphysics. As the decades passed, I would contact various physics departments in prestigious universities to try to interest them in conducting experiments to prove that energy dissolves. In two instances, two PhDs heading the physics department of prestigious universities wrote and told me after reading my dissolving energy thesis that I am in grievous error to believe that energy dissolves. Another physicist told me that I lacked comprehension. Many times my mail to them was returned without being opened (not having a PhD following my name was probably the determining factor). While several physicists did agree with me, their busy schedules only permitted them to give me encouragement.

After Dr. John Wheeler, formerly of Princeton, had moved to Texas, I called him and spoke to him about a test to refute general relativity. Dr. Wheeler advised me to be cautious about how much money I spent on the test. He told me that a North Carolina researcher had spent a small fortune on a test but the scientific world would not honor the test. It is important that you remember the following for your own personal evaluation of Einstein's claims: Dr. Wheeler had proven that the revolution of Mercury's uneven oval is not a relativistic effect but was caused by interplanetary gravitation from Venus, that is, using Isaac Newton's law.

According to Isaac Asimov, in his *Biographical Encyclopedia of Science & Technology*, Galileo Galilei was a very exceptional, talented scholar. His works became a hinge to close old beliefs and open the door to new understanding. The powers that be, both in the secular and religious realms, were miffed and very irritated not only at the manner that Galileo used his deep perceptions, but also his resolve to reveal to all that would listen to the truth found in his discoveries, and how it really pertained to understanding reality.

Pope Paul VI visited Pisa in 1965, and while there he spoke

highly of Galileo. Later, Pope John Paul II stated that Galileo was Italy's greatest astronomer, and Pope John Paul II gave a clear admission to the church's being wrong in the Inquisition and punishment against Galileo. My dissolving energy box is in many ways a repeat of history and is a modern-day Leaning Tower of Pisa experiment.

At this place I am inserting a portion of the *Dissolving Energy Disclosure*,'" copyrighted in 1984. I have edited it to change spelling and to delete redundant paragraphs. Also, much of what was included in the disclosure has been included in the preceding pages therefore I cut the duplication.

Please remember: this work was submitted to The American Physical Society and the American Association For The Advancement Of Science. The envelopes were returned to me unopened. I sent my material to other prestigious organizations, but I never received an answer, or I was told that I lacked comprehension, or was given some other insult. After I posted this section on a Web page, I notified many universities and scientists and asked them to answer the questions. I never received one email, not one. No one ever answered the questions. What other logical answer is there other than they did not want to admit that they have been teaching false science for decades.

I cannot perceive any benefit from naming the scientists that read portions of my work through the decades and refused to answer my questions. I did try in the early 1960s without success to convince President Kennedy's chief scientific advisor Dr. Jerome Baker Weisner that a gravity engine could be built. Also, after I built the energy dissolving box in 1984, I again entreated the US government's wizards through my cousin, a representative in Congress at that time, the Honorable Bill Goodling.

Bill sent the *Dissolving Energy Disclosure*" to the DOE

(Department of Energy) for review. I will include a copy of a letter that was sent to me concerning their analysis.

Ah, yes, readers from across the centuries, I can tell you truly, the mountain that I have climbed has got to be greater than any physical mountain on Earth.

I do not know if my work will be published while I'm living on this Earth. However, in either case, if any of my readers are struggling with God-given projects, processes or and have great difficulty in bringing it into devices reality, remember the government scientist's words to me, Mr. Truett's views have no scientific merit.

Readers: My cousin, Representative Bill Goodling, did the best he could for me by sending my work to DOE. However, the government scientist that received my work in typed pages only saw a small portion of my work and he claimed that my work has no scientific merit. Suppose that I would have added that I prove Einstein's theory of general relativity to be false and I prove astrology to be true? In that case, I may have been in danger of being visited by men in white coats.

It is true that my work (once published) will cause the greatest jolt to organized science it has had in the last two, three or four thousand years—or—how many thousand years would you say?

In our seeming meandering path in physics and astronomy, we have reached a position where creation is pertinent to our development path from dissolving energy, to creating matter.

MAY 3 1985

Department of Energy
Washington, D.C. 20545

May 1, 1985

Honorable Bill Goodling
House of Representatives
Washington, D.C. 20515

Dear Mr. Goodling:

I am writing to follow-up on the April 5, 1985 letter written to
you by Mr. Berton J. Roth of the Department of Energy, with regard
to material submitted to this Department by Mr. Jack R. Truett of
York, Pennsylvania.

I have carefully examined Mr. Truett's disclosure. It describes
his views on a broad range of physical phenomena. Since these views
are very much at odds with extremely well established laws of physics,
such as the law of conservation of energy, Mr. Truett appears to
experience difficulties in having his conjectures recognized by the
scientific community - witness his apparently unsuccessful submissions
to the American Physical Society and to the American Association for
the Advancement of Science.

I am afraid that there is little the Federal Government can do to help.
Under our system, it is not the role of the Government to judge the
validity, or lack thereof, of scientific theories. That function is
left, and very properly so, to the scientific community at large which,
over many years, developed an effective mechanism based on publication
of the new results in refereed journals, presentations at meetings of
learned societies, discussions at seminars, etc. No amount of Government reviewing can substitute for this natural process.

Having said this let me add that as a trained physicist I am personally
convinced that Mr. Truett's views have no scientific merit.

Please feel free to contact me at 353-5995 if I can be of any further
assistance.

Sincerely,

Ryszard Gajewski, Director
Division of Advanced Energy Projects
Office of Basic Energy Sciences, ER-16

Cosmology is the study of the cosmos or the universe.
Beyond Kant-Laplace and the creationists, all matter had
an origin with intelligent design. If our cosmic environment is
the result of a Big Bang, was that the first Big Bang, second, or
third? If at the time of this creation an intelligent design or an

order was established, then a mind either ordered or directed the order. I have a sea shell that I study from time to time. It has very intricate designs formed from a snail's slimy mucous. The shell has a beautiful spiral design with colors woven into the intricate designs. It is a masterpiece of "intelligent design." The two type secretions made by the snail form the intricate designed home for the snail reveals an Infinite Mind with the power to produce intricate designs is present in the universe.

Furthermore, research in parapsychology and reasoning in the realm of the abstract, where metaphysical principles can be cross-examined, continually supports the concept of an omnipresent mind. I have learned to respect our heavenly Father as an Infinite Mind.

All this reasoning and searching to arrive at the same understanding that is given free and concise in the "New Testament" for in it is given in Acts 17:29, "we ought not to think that the Godhead is like unto gold, or silver, or stone."

According to this the Godhead is not something physical. Then consider deeper the expanse of the Godhead as given in I Corinthians 8:6, "there is but one God, the Father of whom are all things, and we in him."

Ah, readers, we in him. We have a physical body that we use to interact with our environment. God has a physical body the physical universe (his Image) that he uses to interact with his created environment through the force and power of the Holy Spirit or Holy Ghost. Christians refer to the mind of the creator as the Godhead. Jews refer to our creator as Hashem; Hindu's refer to our creator as Brahma.

Does the Christian Bible say that God is everywhere in his creation? Consider further:

Acts 17:28 "For in him we live and move, and have our

being."

Therefore, at the time of an explosion between the etheric world (pycnotic mother) and the physical world, order was created and if I may be permitted this conjecture, a fundamental vibration existed from whence all physical creation emerged, that fundamental vibration can be symbolized by a harmonic number.

Whether it is the "OM" sound or what it is can only be approached from theoretical analysis from our low state of consciousness. It is behooving for us to search for a harmonic number that could serve to measure the frequency of electromagnetic radiation's and depict or illustrate the 'rhythm' in these forces.

3569-- number of cyclic permutation having this harmonic quality is found in the number 3569. If we successively add the number 3569, we can illustrate the light wave. Thus 3569 x 10 = 35690 and + 3569 successfully equals: 35690

39259 Violet 800 trillion vibrations per second 42828 46397 49966 53535 57104

Green 600 trillion vibrations per second 60673 64242 67811 71380 74949 Red 400 trillion vibrations per second

If the half of 3569 i.e. 17845 is repeatedly added to 39259 the full spectrum along harmonic development can be shown in 21 steps.

Angstrom Units

21. 1. 392590 Ultra Violet

20. 2. 410435 Violet

19. 3. 428280 Violet

18. 4. 446125 Indigo

17. 5. 463970 Blue Indigo

16. 6. 481815 Blue

15. 7. 499660 Blue
14. 8. 517505 Blue Green
13. 9. 535350 Green
12. 10. 553195 Green
11. 11. 571040 Green
10. 12. 588885 Yellow
9. 13. 606730 Yellow Orange
8. 14. 624575 Orange
7. 15. 642420 Red
6. 16. 660265 Red
5. 17. 678110 Red
4. 18. 695955 Dark Red
3. 19. 713800 Dark Red
2. 20. 731645 Dark Red
1. 21. 749490 Infra Red

Allow me to now open an old notebook. The following disconnected data and notes are from a 30 or so year old notebook. I submit these for the sole purpose of revealing certain systematic logical development that led me to develop a 'theoretical hypothesis' to explain Cosmic Radiation and a heavy particle wave or to paraphrase that term, an electromagnetic wave that is composed of actual physical particles or a physical particle wave.

My work in *My Search for Truth* does not hinge on this hypothesis; it is submitted for 'theoretical conjecture' only.

From My Notebook: Diffraction is a term used in association with light and electron beams. It simply means a breaking of the uniform beam into several beams. Light and electron beams obey laws of electromagnetism for they both are characteristically electromagnetic. One of the aspects of these forces is a phase movement. So, when a beam is broken into several beams, the different beams could affect each other, if they happen to cross in space.

If the beams meet when they are out of phase the individual beams will experience an interference with each other's own structure. If they happen to meet in phase, the two will add together and intensify that particular point where the 'in phase' beams meet. This weakening and strengthening of the crossing areas appears as patterns called diffraction patterns. Huygens' principle of wave propagation, as stated, each point on a wave front may be regarded as a new source of disturbance. The expanding shell like nature of light supports this idea, but this action may not be the cause for the perpetual motion of light.

Since light is independent of the light source, is light self regenerating itself at the end of each phase, renewing itself at each point on the wave front or has the creation of light created a perpetual operating motor that operates in a friction less medium or is the motor in the light wave continually drawing and exuding energy from the ether?

Limits in laws (velocity of light) operating in the cosmos referred to as natural laws are imposed by conditions, determined by forces and counter forces, a meeting, imbrications, reinforcing of electric and magnetic forces.

The Electromagnetic Wave: A changing electric field either expanding or contracting gives rise to a changing magnetic field inversely related so as the one type is expanding, the other is contracting in opposite direction either clockwise or counterclockwise and vice versa. The inverse relationship conserves the energy and provides a directional propagation by reason of the fact that electric field does not close on itself (on its own movement); rather, it provides a propelling action from a spiral section.

I wish to interrupt at this point from reading my notes and make a point clear.

From my notes pertaining to the propagation of electromagnetic waves, it is important to know that a concept pertaining to clock wise and counterclockwise movements is what I believe to be original. I do not recall reading in any other physicists work any reference to clockwise and counterclockwise movements within the wave.

This concept came from or through my mind as far as I know and it is a concept that I still hold. Although I have become more conservative in my old age and do not wish to submit this as part of a generalized theory. Time will surely reveal more truths.

I became very curious about clockwise and counterclockwise motions in nature during the middle sixties and I wrote to Dr. George Gamow (1904–1968) another renowned scientist working at the University of Colorado to compare notes with him. When I wrote Dr. Gamow, he was very busy then went on leave from the university because of a health problem. After writing the second time, another faculty member answered my letter and informed me that he knew nothing of Dr. Gamow's interest and also he knew nothing of Dr. Gamow's interest in certain clockwise and counterclockwise primitive life forms that mated with only other life forms with the same left or right hand revolving motion. He also informed me that Dr. Gamow took up residence in another environment in 1968. However, he simply told me that Dr. Gamow had died.

Dr. Gamow had a son that was referred to in Dr.Gamow's book *The Birth and Death of Our Sun* as a son, "that would sooner be a cowboy" grew into a man to become a scientist in his own right.

I also do not remember reading anywhere that it provides a directional propagation by reason of the fact that electric field does not close on itself (on its own movement); rather, it provides a propelling action from spiral sections provides a propelling action from a spiral section.

Let us now return to my notes from an old notebook.

Ethons of the ether should not be considered just to be useful to account for energy dissipation, neutrinos, Photons and every other intellectual bothering demonstration of nature. Ethons of the ether are real and are necessary to unite the wave and particle characteristics of electromagnetic radiation, to the quantum principle and the conservation principle from the perspective that Ethons are the building blocks of all creation.

If Ethons could be destroyed, then the universe would be dissolving, our minds would dissolve and God would dissolve. Ethons cannot be destroyed.


**HOW ELECTROMAGNETIC FORCES AFFECT THE ELECTRON.**
Electron diffraction and interference patterns, electrons wave characteristics. An electron is a wave or wavelet where the phase is formed in a circle as a caterpillar rolling into a ball with the electric field on the outside of the electron.

Test To Be Made - Form Electrons From Light By A Circular System Of Reflectors.

I had forgotten about this idea that I had conceived to form electrons from a curved circular and reflecting device. (Picture in your mind the inside wall of a doughnut). Since we now have almost perfect mirrors that was given to us through fiber optics, it would be interesting to make a circular path for light that would be open in the middle, yet very close with a place for light to enter and enter a closed circle, or another variation is a spiral with decreasing diameter until the spiral becomes very close and closes on itself.

Also—the glass used to make the fiber optic mirror could be doped with gold, platinum or copper since the field in matter is an important factor in forming electrons that have a negative charge. Certainly if greater light energy would be directed to the closed circle, the light would not break up as in the box experiment where I designed it with glass to diffract or to break up and dissolve the light, instead, mirrors of near perfect reflection as fiber optic quality formed in a curved circle should precipitate particles. Hmmm, I wonder how easy it would be to make a low-energy particle beam. This conjecture was not part of my notes. This aspect of forming electrons from a reflective closed system reminds me to mention something about the so-called cold fusion.

Research scientists and some from MIT are claiming the Pons-Fleishman experiment is phony because it does not obey the laws of nature. Perhaps these are some of the people that stated that, according to the laws of aerodynamics, a bumblebee cannot fly. However, the bumble bee is oblivious to the laws of aerodynamics, and flies with ease.

These are the same people that are saying that general relativity is a true and valid theory, and astrology is a phony science because the planets do not affect human thought or behavior.

We researchers have not discovered all the secrets in nature. There may be an unknown process at work with the gaining heat from the special electrolysis. I suspect that more than one condition is responsible for the heat gain. I suspect that it can be controlled, but knowing what forces are at work is necessary to greatly shorten research time.

Since all or some heavy water was electrolyzed, it has been claimed that some type of fusion occurred that caused the increase of temperature.

I have not read every article about the 1989 Stanley Pons-Martin Fleishman experiment, and therefore I do not know the claimed cause for the rise in temperature. However, since heavy water was involved, it raises the question about freed neutrons being the source for the rise in temperature. Neutrons are unstable when not in a nucleus, and transmute to one proton and one electron. You know what that is, don't you? It is a hydrogen atom.

If a neutron transforms to a hydrogen atom, some heat could be a byproduct. In 1958, Dr. Samuel Cohen, while researching at the Lawrence Livermore National Laboratory, invented the neutron bomb. Its basic atomic component is tritium, and that is, hydrogen with two neutrons in the nucleus. This bomb, when exploded, is radioactive.

However, hydrogen with one neutron in the nucleus is deuterium. Since there was supposed to be all or some molecules of heavy water with the Pons-Fleishman experiment, that aspect comes to the front as one possible source of heat.

The research scientists at MIT that claim the Pons-Fleishman experiment is phony because it does not conform to the laws of nature. They are of the same breed that said to me, energy does not dissolve, astrology is a false science, and general relativity is true. Also readers, I don't suspect these researchers ever tried to make electricity from gravity at Grants Pass, or were real serious about replicating the possible heat affect from electrolysis using heavy water.

In analyzing this phenomenon, it is necessary, to consider the geometry of the porosity in the plates or rods, as a possible factor in developing heat. However, I am inclined towards the possibility of freed neutrons transforming into a hydrogen atom, and in the process creating heat. Another factor that is important to point out, in the Pons-Fleishman experiment; it was claimed that particles have been formed that are way above the mass of electrons.

If this is so, it supports my suspicion that extra neutrons transformed into hydrogen, and heavy electrons. This would definitely generate heat. One more possibility should be considered. I do not know what type of electrolyte was used with the water when separating water into hydrogen and oxygen. However, if a neutron or neutrons were freed and dislodged an atomic electron, this would bare a proton. That would generate heat and electrons, for protons are not stable without an electron.

Cold fusion is a misnomer for their experiment because in the process, atoms are not split. I grant this idea that I have presented here, freely to any university. Since I already know how to use a catalyst in electrolysis to greatly facilitate the water separation, I strongly suspect a process of electrolysis using heavy water will generate heat. If the water being electrolyzed is not pure heavy water, then how much heavy water is a factor that should be determined in numerous experiments?

This thinking causes me to reflect on the atomic pile, and Enrico Fermi. There may be several variables involved with, let's call it "Heavy Water Electrolysis." If electrolysis with heavy water does generate heat, it would require an engineering project to make an electrolysis vessel that can generate steam, capture the steam and use it to generate electricity. However, heavy water is expensive to process from ordinary water, and it requires electricity to separate water into its atomic parts. If replicated, I don't expect "Heavy Water Electrolysis" to ever become a commercial venture.

So, I parry the theoretical conjecture, can electrons be formed by separating the heavy water molecule to hydrogen and oxygen? If so, then through a process of freeing extra neutrons, could hydrogen and sub atomic particles be the transmuted particles. If this be true, could the freed neutrons reach a state where heat is generated and stable particles remain? Only research will determine the truth.

For over 35 years I thought it to be very reasonable and very, very possible for electrons to experience an energy change where they become heavier stable particles. My initial work in this line of theoretical reasoning began with a study of Cosmic Radiation' (particles radiated by the Sun) commonly known as 'Cosmic Showers'. It is phenomena studied and speculated upon by Arthur Holly Compton. Compton was not the first to study 'Cosmic Radiation,' for Robert Andrews Millikan had preceded him in this research. Compton did extensive research, traveling a lot over the Earth to determine if the Earth's magnetic field had an influence on 'Cosmic Radiation.' I was very concerned with the nature of 'Cosmic Radiation,' for I wanted to know their place of origin, how they were formed, why they are as they are and what their eventual destination is.

I offer the following theoretical conjectures in larger type for future generations to consider when studying my work. It is important to remember that this perspective came from my mind as I was working as a baker raising a family and studying from some books that I had bought from a used book store.

*Cosmic Radiation* is a term that encompasses three specific types of particle radiation, Alpha, Gamma and Beta.

1.  Alpha - two packets - a proton and a neutron in each

2. Gamma - two packets - an electron and another particle of equal mass to the electron but with an opposite charge therefore called a positron

3. Beta - two electrons in close proximity

Beta rays could be several sets of electron groups. *Cosmic Showers* as a term was developed as a result of 'Cosmic Radiation' entering the Earth's atmosphere. When this occurs the field in matter (the gases of the atmosphere) interacts with the particles and a breaking of the particle formations occur, and a condition results that has been named 'Secondary Cosmic Showers.'

'Cosmic Showers' always occur during Sun spot activity, while the Sun's magnetic field reverses. It was known that the Sun always spewed out these energy particles during a period of intense solar storms. Whenever this happened on the Sun, it appeared as 'Sun spots' here on the Earth. It was calculated to be an eleven-year cycle. (some have speculated that it was the studying of Sun spots by looking at the Sun without eye protection that contributed to Galileo becoming blind in his last years of his Earthly life).

Because the Sun has such a powerful gravity field in its central core, we can expect the lightest atoms (hydrogen) would be found on the surface. Of course we know this for hydrogen being transmuted to helium is the fusion furnace. If we could slice a wedge from the Sun and examine the interior, its logical to find that a similar type of onion shell structure exists where all the atoms on the periodic table are present in a state of plasma*, but - in an orderly sequence beginning with hydrogen on the outside and blending into each other as hydrogen is followed by helium then lithium, beryllium, baron, carbon, nitrogen and so on with the heaviest atoms in the core.

Plasma is an atomic state of matter where electrons have been stripped from the atomic structure by intense heat. It was named by Irving Langmiur (1881–1957) an American chemical genius. Great solar storms would cause interior atoms to be brought to the surface, as proven by the dark lines found in spectroscopic analysis first discovered by William Hyde Wollaston (1766–1828) then to greater precision by the renowned scientist Joseph von Fraunhofer (1787–1826).

1. Remember the diffraction aspect of light as it interacts with itself and cancels some part of the wave structure and reinforces or strengthens other parts of the wave?

2. Remember also that light travels by a wave action of expanding shells and waves are the affect of actions and reactions of component parts operating within a system. Also - a component energy particle in the light wave named a corpuscle by Isaac Newton and over two hundred years later named a photon by Dr. Gilbert Lewis has never been found.

However, take careful note, which does not mean that Photons do not exist for concentrations of energy in the wave structure do exist. However, they do not have sufficient energy in their composition to impart any mass to their individual identity therefore a photon is not a stable particle to exist alone or independently. As these energy concentrations or component parts of the operating system cause continuous changes by their actions and reactions within the system, the photons are also experiencing changes.

3. Because - Remember This Too - A changing electric field either expanding or contracting gives rise to a changing magnetic

field inversely related so as the one type is expanding, the other is contracting in opposite direction either clock wise or counter clock wise and vice versa.

But again take careful note - the expanding and contracting of magnetic and electric fields, IS a building and collapsing of magnetic and electric fields. It is my understanding and I submit to physical science, at the point where the building and collapsing field occur, is the point where the light wave experiences an absorption of Ethons or an exudation of ethons, depending (of course) whether the light wave is advancing from the red end to the violet end or from the violet end to the red end.

The inverse relationship of the magnetic and electric fields acts as a transformer while preserving the system and provides a directional propagation by reason of the fact that electric field does not close on itself (on its own movement) rather, it provides a propelling action from spiral sections of the electric field, and since it does not have a 'perpetual motion' motor, the light wave absorbs and exudes energy from the ether, and this is nature's "perpetual motion motor."

4. Remember also the Huygen's principle of wave propagation, as stated, each point on a wave front may be regarded as a new source of disturbance. The strength of this comes from the expanding nature of light.

We must keep in mind the fact that electric field and magnetic field are two interrelated forces and are inversely related as building and collapsing fields. We will consider the Huygen's principle by analysis very shortly.

5. We also know that in all electromagnetic radiation the wave length varies inversely with its frequency. The forward

movement remains the same or constant, therefore, the forward movement is independent of the frequency, thus, either the wave advances by many small steps or few large steps.

6. Remember also that different colors in light have different chemical affects upon matter and (as music) have differing psychological affects upon life forms.

I could not perceive where photons and any linking energy chains could vibrate at 390 trillion cycles a second and have one psychological affect on life forms then by increasing the oscillations to 570 or 760 trillion cps have vastly different psychological effects.

I learned from my searches into forces that affect consciousness that behavioral changes also are accompanied with chemical changes. In the cases of classic schizophrenics when ever schizoid or as I refer to the commanding alter ego that usurps the 'seat of consciousness' schistic personages acquires identity, chemical changes occur in the body.

Therefore, after the comparison to biological life where changes are associated with chemical changes has been given, we return to similar properties of light. Since light does have different chemical effects upon matter and has differing psychological affects upon life forms let us acknowledge that it is true and do a little detective work. We know the upper violet rays of light has a bleaching power and will cause melanoma or 'skin cancer.'

We also know the lower infrared light will increase 'thermal action' or generate heat. We know that 'green light' has a beneficial effect on photosynthesis probably because of its affinity with chlorophyll.

Since colored lights affect 'human moods,' a strong case has been established that not only can light cause 'melanoma', but also can influence the production of hormones in the human body. Remember, human behavior is tied to body chemistry.

Therefore, not only does the light wave have inherent vibratory differences from one end of the spectrum to the other, its chemical differences if following natures patterns, probably comes from particle geometry.

Consider: What is the difference in the elements? It is a vast difference in chemical differences and physical properties. True, but only three particles are used; it is the geometry and the number of the three particles that produces such vast differences.

Do we have to stress our analytical thinking to submit the following proposition: The Light Wave's Inner Workings Have Differing Geometry Inherent To Each Color. That is not a bold or radical departure from what is known about light, however, if science denies the pycnotic theory, and denies the existence of the ether, and holds tenaciously to the 'conservation of energy principle' instead of accepting that energy dissolves, then this speculation into the 'realm of the abstract reasoning' will not advance science.

Regardless of how strenuously the 'old guard' tries to hold to old and out dated beliefs the facts are abundantly clear, light has a changing electric field either expanding or contracting that gives rise to a changing magnetic field inversely related so as the one type is expanding, the other is contracting in opposite direction. The inverse relationship conserves the system and provides a propelling action. While the light wave generates its wave, 'inner workings' produce changes to the quanta of energy in photons.

Now, I ask you - are the two inversely related expanding and contracting forces or building and collapsing fields completely independent of each other or, are they structured together at a root?

I asked, are the building and collapsing fields structured together at a root?

It Is Possible For The Light Wave To Have A Root. Assuming this to be true, It is logical for the electric and magnetic forces to expand and contract while connected to a root stem. This stem could revolve in a clockwise or counterclockwise direction. It could be possible for the revolving stem to cause the electric and magnetic forces to build and collapse, or expand and contract.

Therefore, the root stem would be the heart and Soul of the light wave and when this is broken the light wave ceases to exist. This could explain how the light wave can continue with sections of it removed as demonstrated by Fraunhofer lines and the canceling and intensifying of the light wave when it contacts other light waves and interacts in a manner to cause the diffraction patterns.

Arthur Holly Compton's work revealed that photons are packets, or aggregations of energy that form a nebulous particle form, within the light wave structure.

Therefore it follows: As the light wave generates its wave 'inner workings' it produces changes to the energy aggregates or photons. But in addition, the oscillations contain a quanta of energy that is dependent upon its frequency; therefore, the light wave is continually absorbing Ethons as it advances from red to violet and exuding Ethons as it advances from violet to red. Therefore scholars, not only is the advancing light wave increasing and decreasing energy as it advances, the geometry of the constituting tenuous particles is changing.

Consider the following: We use the term 'component parts

operating within a wave system' but let us again review this principle. We know that light expands or advances by a succession of spherical shells. For example: if we make an iron or steel sphere equipped with an internal knocker that we can activate by remote control and cause the knocker to hit the interior wall, the sphere will vibrate all over its exterior spherical surface. A ringing sound could then be heard.

Sound waves would compress and rarefy the air all around the sphere in an expanding shell pattern. Sounds would be able to be heard and detected with equipment from innumerable positions surrounding the sphere. Similarly, if we use a spherical fluorescent light bulb illuminated by radio waves, light will radiate from the light bulb with a similar expanding shell pattern.

In addition to the Huygens principle, the light wave has its own motivity and draws energy from the ether and exudes energy to the ether as it advances.

I have acknowledged this principle to be operating not only in the light wave, but in all electromagnetic forces.

# The component parts in the light wave system are continually absorbing and exuding ethons as the wave advances.

Consider the following: If you then can perceive that the light wave is not only moving through the ether but is absorbing ether or more correctly ethons that is converted into energy as it takes its place in the wave system and exuding ethons as the light wave advances from the shortest vibrations to the longest (violet to red) we are in agreement.

Thus, the light wave 'breathes' or absorbs and exudes ethons from the ether and that is exactly the principle of pycnosis.

It is imperative to illustrate this principle by a test. Sunlight for this experiment will be fine. The test is simple; sunlight is to be separated into the full spectrum by a prism. The divided light is to be directed into an opening about two inches wide and twelve inches long. Where the light makes contact with a flat back behind the long slot, it should be covered with solar cells to convert the light to electrical energy.

The spectral colors of the prism must be slowly turned to allow each color to shine through the slit, and each color will be able to be converted into electrical energy. The electrical energy from each color must be measured as it is transmitted to a resistance, small light bulbs. In the far right end is to be red color. It could be violet but a reverse order will be created after readings are taken with red on the far right. As the light is directed onto the photo cells, each color converts light to electricity and the readings show the red end to have the least amount of energy.

As readings are made, each color going from red toward violet is producing a greater amount of electrical energy until we reach the violet end where the reading of electrical energy is about twice as much as the red end. Therefore, all along the line from red to violet there has been an increase of energy in the light wave which is proven by direct measurements of increased electrical energy. Therefore, the light wave was absorbing energy as it advanced from red to violet. Now comes the great grip on truth and reality.

There is a connective operating system within the light wave. As the intricate wave advances, the energies that operate within the wave system and the Photons that advance with the wave, were either absorbing or exuding energy.

From the red end to the violet end, the light wave was absorbing Ethons from the ether as the wave advanced. If we take readings from the violet end first and take readings all along the spectrum to the red end, we can prove that the light wave exudes energy, because as the light wave advance in free space, its advance is a continual repeat from red to violet then back to red and back to violet and continuously repeating the cycle.

Conclusion: the light wave absorbs and exudes Ethons from the ether as it advances and that is exactly the principle of pycnosis. But the absorbing and exuding energy inherent to the light wave is also an operating principle in all electromagnetic radiations.

Since the light wave has 'component parts operating within a wave system' and since it vibrates at between 400 and 800 trillion times a second, we would never be able to see the 'operating system.' However - since the known laws of nature reveal a 'harmony' and repeating patterns, it creates a background framework where we are able to speculate based upon reasonable premises. For example: we know 'black holes' have such a powerful gravity field that light is compressed and cannot escape. We also know that gravity is intense at the core of a star. But, even this has degrees or a gradient for there are 'Big Reds' 'White Dwarfs" and by comparison the center of our galaxy.

With the consideration of the vast differences in 'Gravity's Force,' it would be possible for an order based upon a gradient where the most intense gravity has the power to compress light waves until a "higher order of creative forces" creates—particles?

Please keep your concentration for we are now approaching a broader understanding concerning the origin of matter. Since light can be caused to reinforce with itself when conditions create interference patterns within the gravity field of the Earth, isn't it very logical to consider the possibility that 'intense gravity' at the core of the Sun could form particles from compressed light waves?

Since I have led you this far, allow me to submit a reasonable hypothesis for the origin of cosmic rays. I believe it is very possible for "Alpha Rays" "Gamma Rays" and "Beta Rays" to be formed in one continuous wave system at the center of the Sun and are 'spewed out' during violent storms on the Sun that we see as Sun spots.

It is also very logical to assume that "component parts operating within the wave system" are not all stable particles. Therefore - only the stable particles are ever discovered here on the Earth, for the unstable particles become disintegrated when the wave structure is broken.

From these known facts and 'theoretical conjectures,' based upon extensions of known phenomena, a higher order of creative forces could compress light waves into a "heavy particle wave," where the tenuous photon particles become physical particles.

Allow me to clarify myself concerning the phrase a higher order of creative forces could compress light into a "heavy particle wave" where the particles are physical particles.

I believe there is a principle in nature operating that can be referred to as "the conservation of identity principle."

While I have never read of this anywhere and probably you haven't either, therefore, you will again have to follow my line of reasoning. Please consider the following: within the center of a Sun or a 'black hole' where gravity is tremendous, light waves (and other electromagnetic radiation above and below the light wave) are experiencing crushing forces that could condense or compress matter and electromagnetic forces into a dense condition. (Arthur Holly Compton was the first to speculate about this possibility, for he stated, "If waves be particles, why not particles be waves")

However, within a "black hole" or the center of a Sun, protons and neutrons are stable particles and resist being broken into energy fragments. I also believe that electromagnetic energies are stable, albeit, very tenuous. Therefore, it is very logical to consider their identity is preserved by absorbing more energy.

Thus - the packets of energy within a light wave absorb ethons that become energy, to retain identity and become particles. I mean physical particles, protons, neutrons plus the electrical particles, electrons and positrons. Please follow the following explanation carefully.

The following theoretical concept concerning the creation of particles from compressed light was copyrighted by this author in 1963.

Understand, the Alpha rays occupy the same position as the violet end of the light wave, the Gamma rays occupy the same position as the green in the light wave and the Beta rays occupy the same position as the red end of the light wave.

Whether from "black holes" or from the center of a star, the above 'theoretical concept' should be considered seriously within Grand Unified Theories of physically particles (GUT's). Protons, neutrons, neutrinos, electrons, positrons, mesons, plus all energy particles (also if you desire quarks and maybe barks) for physical particles are not hatched, they are compressed into being.

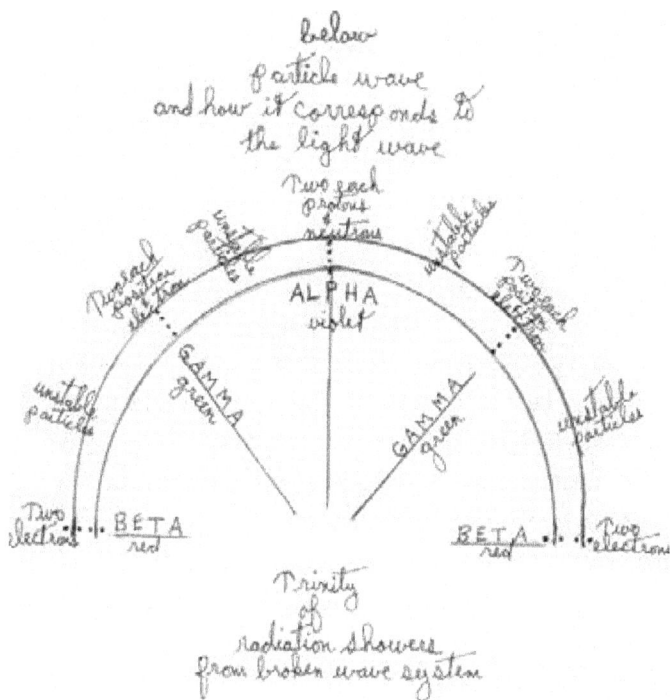

below
particle wave
and how it corresponds to
the light wave

Two each
protons
&
neutrons

ALPHA
violet

GAMMA
green

GAMMA
green

BETA
red

BETA
red

Two
electrons

Two
electrons

unstable
particles

unstable
particles

Trinity
of
radiation showers
from broken wave system

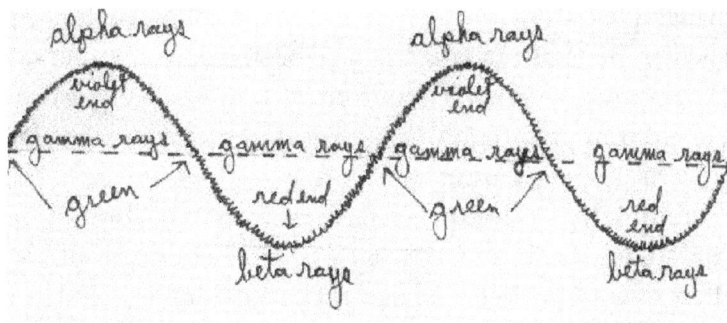

alpha rays

alpha rays

violet
end

violet
end

gamma rays

gamma rays

gamma rays

gamma rays

green

red end

green

red
end

beta rays

beta rays

Please consider the following: the energy particles within a light wave (Photons plus every and all other packets of energy in the light wave) are propelled forward by an etheric force and are continually transformed in energy density and frequency as the light wave advances. Similarly, an electric current advances by reason of the electrons being propelled forward by the electromotive force. Thus it is very reasonable to associate the nature of the light wave and electricity in the relationship where actual physical particles are created. If waves be particles why can't particles be waves? That may not be an exact quote from Arthur Holly Compton.

Light is an electromagnetic force as        electromagnetism is an electromagnetic force that drives electrons along in an electrical current. Electromagnetism or EMF is measured by units named voltage, whereas the flow of electrons is measured in units named amperage. Direct current, i.e. electricity that flows from a battery, generator or dynamo, flows as a continuous train of flowing electrons in one direction whereas Alternating current i.e. electricity created by an alternator flows in continuous surges.

The surges enable the turning off and on nature to cause a building and collapsing field within a transformer, thereby changing electrons to EMF or EMF to electrons dependent whether the transformer is a step up or step down transformer. So, electromagnetism is transformed into electron particles and electron particles are transformed into electromagnetism or electromagnetic forces.

Similarly, an advancing light wave is transforming its Photons into various energy content particles from the infrared end to the ultra violet end and back to the infrared end as it moves through its repetitious seven-step cycle.

These known conditions of electricity and the light wave leads to the "heavy particle wave" where stable particles of energy (electrons and electrons plus positrons) are continually transformed in energy density and frequency as they are propelled by a more powerful force than that which propels light and electricity until the end of the cycle where the energy composition is about 1,844 times greater where two packets of energy particles become physical particles (one proton and one neutron in each packet.)

Once reaching the end of the cycle the down side of the cycle commences as energy is exuded, frequency lengthens until the place on the wave where four electrons occupies the place at the opposite end of the cycle. Therefore, unlike the light wave which is composed of tenuous particles that are not stable outside of the wave system, i am submitting the "heavy particle wave" enters the atmosphere of the Earth where the wave structure is broken, and the particles being stable are classified as cosmic showers.

I consider the above concept much more reasonable in theoretical conjecture than the radical idea postulating that planets revolve in an "inherent curvature of space" due to space's "non linearity," revolving in a "Space Time Continuum" "by inertia"—"taking the shortest possible path in a warped and deformed space" where "gravity is a condition of space in the vicinity of matter." Inertia causes the planets to move around the Sun!! Secular scientists: It is time to rise above delusions and ask God for greater understanding about his universe.

Gravity *is* a condition of space and that is not new for mass radiates gravity into space. What *is* new is the new knowledge of an Aust that is created by every rotating sphere. The condition of gravity in space or its properties was carefully explained to humanity through the minds of Kepler, Newton, Cavendish, and now the Aust and the reintroduction of the pycnosis concept is given through Truett.

All physical particles radiate gravity waves that exert a

pulling force that is centralized in the core of each particle. At the core of every particle, every particle is absorbing ethons from the ether to balance the energy that is being radiated. What has not been discovered about gravity is how we can cause the pull force to be reversed to a push force and use that to propel space ships. Yet, space is not complicated either; space is the ability to hold something. Perfect space is a vacuum, although ether is everywhere.

Scholars of science: Chapter 1 of *My Search for Truth* Volume I was devoted chiefly to energy and the solar system.

A Change Of Focus Follows

There are physical phenomena that are very destructive and have been most fearsome to man for as long as man has been on this Earth. These fearsome events of nature are Earth Quakes and Volcanoes. My research into volcanology and diastrophism has yielded some new discoveries about the nature of these two fearsome, cataclysmic events. We will close this volume with an in-depth study of these phenomena.

Volcanology AHD - "The scientific study of volcanoes and volcanic phenomena." Diastrophism AHD - "The process of deformation by which the major features of the Earth's crust, including continents, mountains, ocean beds, folds, and faults, are formed."

# CHAPTER II

# EARTHQUAKES, VOLCANOES, AND HURRICANES

This chapter reveals the reason for earthquakes, the reason for volcanoes and how to control them, and an expanded understanding about what causes hurricanes and how to control them.

Earthquakes: I submit to all my readers, to the world now and into the future until the end time of man on this Earth, the reasons for earthquakes.

EARTHquakes are divided into two types. Type one earthquake or "fault line earthquake" is the most devastating and is caused by the movement of the Earth's tectonic plates where tectonic plates scrape against each other as one moves faster or moves on top of or beneath the other.

Two factors or forces cause "fault line" earthquakes. These are: Centrifugal force from the rotating Earth and gravitational attraction from the Moon, Mars, and Jupiter.

Please take careful note of the given graphic.

It is plotted using heliocentric co-ordinates, meaning the location of the planets are plotted according to their exact position, not as they would appear from a geocentric perceptive.

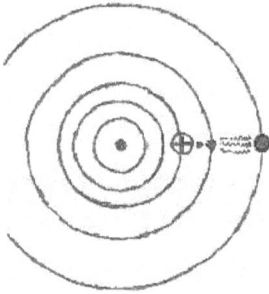

**I SUBMIT:**

The attractive force of the Moon, Mars, and Jupiter is responsible for slowing the rotation of the Earth.

This causes a second to be added by the time keepers (leap seconds) to keep a day exactly 24 hours.

In addition, this is the earthquake alignment.

The earthquake alignment graphic above illustrates the planetary configuration that will cause "fault line earthquakes", volcanoes and exceptionally high tides. It is the *only* planetary configuration that will induce the phenomena given. This configuration is a conjunction of the Earth, Moon, Mars, and Jupiter being most powerful when the alignment is at full moon, when the Earth is at aphelion and the Moon is at perigee.

I have naturally named this alignment the "earthquake alignment" with the secondary term "high tide alignment." During this alignment, as given, Venice, Italy, may be super flooded (other factors are also involved with high tides in Venice.)

If an uncontrolled hurricane would strike the American eastern coast during this alignment, when the tides are exceptionally high, the storm surges would be super destructive.

During the destructive hurricane to the Florida Keys on Labor Day September 2, 1935, Venus and Earth were in alignment. Although, the 200 mph winds were the chief factor that caused such devastation in the Keys.

In the future, in addition to the Earth warming, gravitational forces from Venus and Earth in conjunction, plus the attractive force of the Moon, Mars, and Jupiter when aligned, are factors that meteorologists will have to include in calculating the severity of hurricanes. However if the solution to control hurricanes is followed, hurricanes can be deflected away from the American coast line.

The destructive hurricane named Katrina that made landfall on August 29, 2,005 at New Orleans at 11:00 AM occurred while highest tide was approaching being only one hour and 56 minutes to high point. That push of high tide rising made the storm surges very destructive to a city that is twenty feet below sea level. I personally watched the weather channel and I did not see any special focus made to the high tide rising factor.

Type two earthquake is a " Moho Mantle caused" earth tremor. These tremors are created by the mantles of the earth moving to restructure its cracked compositions. Moho Mantel caused earthquakes were not frequent before huge dams on rivers were built. After the Hoover Dam, Aswan Dam, and the Three Gorges Dam were built and "fracking" was used for oil exploration, "Moho Mantle caused " earth tremors increased. The reason for this is caused by the huge amount of water pushing down on these regions with some water being pulled into cracks in the mantle and thereby causing the mantle to move to restructure its composition. When this occurs tremors are felt on the surface and are not yet classified as " Moho Mantle caused" earth tremors or type two earthquakes. This work differentiates between type one and type two earthquakes.

**VOLCANOES**

In addition to various reference books that are devoted to vol-canology, there is a lot of information about volcanoes on the Web. I have found Google.com to have the greatest listings of sources that explains volcanoes; even NASA has a site on Google. com that goes into great detail about the nature of volcanoes. It should not be a surprise to my readers when I state that I disagree in major differences with organized science as to the nature of volcanoes. Also,

I have not found any of these prestigious sources for information to explain how to control volcanoes. These prestigious organizations would not read my work presented to them that proves astrology and proves that general relativity is a false theory—so—my work that presents a method to control volcanoes and hurricanes would be just as hilarious to them.

In addition, I haven't found from any reference source to my satisfaction, a carefully described *reason* for earthquakes, volcanoes, hurricanes, leap seconds and the shifting of the Earth's magnetic poles. Certain references attribute gravitational force to be one cause for maintaining the Earth's hot central core. However, the Earth was once a molten glob of Sun matter, and has been cooling for over four billion years. The Earth's Moon was once plasma then molten and it has cooled and is now a solid mass, without a hot core.

Not only did the Earth's Moon completely cool, other moons in the solar system excepting perhaps the three largest moons, Triton, Ganymede, and Titan, they may have a molten core. Ganymede and Titan are larger than the planets Mercury and Pluto.

The idea that purports the Earth's central core is hot due to the force of gravitation, is false. In time, (millions of years) the Earth will be completely cooled, and the central core will be cool as is the Earth's Moon. Also, in this chapter, the new understanding about Tectonics is incorporated into a general explanation for earthquakes and volcanoes. I ask all my readers to access these sources, NASA and Google.com and compare what is given to the following explanations.

When you read the following paragraph about the Earth's cyclic crustal movements, you will also learn what causes volcanoes, how to reduce sporadic volcanic eruptions and how to reduce El Niño affects.

As the Moon revolves around the Earth within the Earth's Aust, it passes Mars and Jupiter every revolution and it also passes Mars and Jupiter when they are conjuncted (Earthquake Alignment.)

During the Moon's conjunction with Mars, Jupiter, and the Mars-Jupiter conjunction, great gravitational force pulls on the Moon and due to this pulling force, I submit, the greatest force is exerted that very slightly slows the Earth's rotational motion.

In addition, I submit—the combined pulling force of Mars and Jupiter pulls the Moon farther from the Earth. According to NASA data, the distance being, an *average of one and a half inches a year to three inches a year.* Astronauts on the Apollo missions 11, 14, and 15 placed reflectors on the Moon specifically to be used in measuring the distance between the Earth and the Moon. Laser light was later beamed to the placed reflectors from several astronomical facilities at different times of the year, including the Lick and McDonald observatories.

Whether the receding distance be an average three inches a year or less, NASA scientists did a great job at discovering how much the Moon is receding but failed to understand why.

I have never read where another scientist offered an explanation as to *why* the Moon is slowly moving away from the Earth, so this explanation that you are reading in this work, I believe is another original in this work.

No one needs to be a rocket scientist to see that this alignment shown where Jupiter, Mars and the Moon can pull the Moon slightly away from the Earth. In addition, when the Moon, Mars and Jupiter are all in conjunction with the Earth this alignment is capable of causing the shifting of the Earth's magnetic poles over long periods of time and for producing earthquakes and volcanoes through fault lines. The "Earthquake Alignment" that is presented in this work *should be* taken very seriously.

Consider the following: On November 10, 1977, Jupiter, Mars, and the Moon were in conjunction with the Earth and shortly before November 10, on September 13, Kilauea Volcano on Hawaii had one of its biggest eruptions that lasted continuously for 18 days. In 1983, Kilauea again erupted, and due to a shifting within the Earth, that caused a pipe to remain open to the magma, Kilauea has erupted continuously to this date as I edit this volume in January of 1999.

Without a doubt, the continual eruption of Kilauea has been a great blessing to humanity. The release of Magma pressure from Kilauea's continual eruption has prevented more explosive eruptions from occurring, in other parts of the world.

On June 12, 1991, Mount Pinatubo in the Philippines erupted with a terrific explosion. Following that on October 6, 1996, Iceland experienced a tremendous volcanic eruption that lasted from the sixth to the twelfth. In addition to these Mt. St. Helens, Montserrat, and many others throughout the world have had powerful eruptions.

Now pay attention closely, for in addition to November 10, 1977 when Jupiter, Mars, and the Moon were in conjunction, Jupiter, Mars, and the Moon will *again* be in conjunction with the Earth on August 17, 2013. For astronomers that doubt my claim, I entreat all astronomical observatories to recheck the positions of Jupiter, Mars, and the Moon on the given date.

**CONSIDER MY ANALYSIS**

After about a half a billion years or more after the Earth was created, a huge land mass or continent was formed by continual volcanic eruptions.

Two forces caused the huge mass to break into separate land masses. First: centrifugal force upon the uneven shaped rotating Earth exerted a powerful force that contributed to the land mass breaking into separate land masses; these are now called tectonic plates.

Secondly: the influence of gravitational forces from the Moon, Mars, and Jupiter, caused and still causes the softer layers beneath the crust to move.

These continual forces upon the Earth cause crustal movements and this phenomenon is known as earthquakes.

To understand my concept in finer detail considers the following. From the formation of the Earth when it was a glob of Sun matter, more than four billion years ago, its shape has changed into a five-layered sphere that is bulged (pushed out) at the equator region from centrifugal force. The diameter at the equator is 26.6 miles greater than regions above and below the equator, but due to the Earth's oblateness, the equator diameter is about 84 miles greater than the diameter from pole to pole. In addition, the Earth is pushed out at the magnetic north pole 50 miles and pushed in 50 miles at the magnetic South Pole.

A drawing was given on these pages to depict the Earth's shape. The pushed-out and pushed-in region of the poles is a natural condition from a powerful flowing magnetic force. Einstein was mystified by the Earth's magnetic field and referred to it as one of the last remaining mysteries. The true mystery is how organized science could continue to teach Einstein's opinions as facts. Please take careful note: these polar pushed out and pushed in regions on the Earth and also on the Moon is caused by what I am presenting in this work as a flow of magnetic force into the south pole and out of the north pole.

........................................................................

# The earth's physical nature, the shape of the earth's interior, has layers as the shape of an onion. However, the earth has only five layers, and as these layers goes deeper towards the center, the heat increases.

........................................................................

In regard to the Earth's spherical five-layered structure they are as follows: From the surface of the Earth, taken at the equator, to the center is about 3,900 miles. That is roughly one half the diameter of the Earth, at zero line of latitude.

1. The surface of the Earth is the outer skin of the *crust* which ranges in thickness from about three miles under the oceans to 25 miles and 35 miles under the highest mountains.

2. Under the outer crust is a *crust base* named the *Mohorovicic discontinuity* (referred to as the *Moho*) which averages about five miles in depth under the oceans to twenty miles in depth beneath the continents.

The Moho is an interface between the outer crust and the *mantle*. This layer was named after a Croatian geophysicist,

Andrija Mohoroviâiä (1857–1936.)

3. The third layer is the mantle, which according to the divisions that have been made constitutes the greatest composition. The mantle's composition ranges to solid matter where it blends with the Moho or lower crust region to near molten and molten where it blends with the magma and is about 1,800 miles thick.

4. The *molten magma* is about 1,400 miles thick and is seen on the surface of the Earth as flowing lava from a volcano.

5. Beneath the magma is the *solid spherical core* which has a diameter of about 745.6 miles. The tremendous gravitational force that pulls the crust, moho, mantle, and the magma towards the center keeps the core in a solidified state.

Just as we must recognize that earthquakes and volcanoes have a physical force responsible for the phenomena, we also must attribute a physical force to be responsible for causing the shifting of the Earth's magnetic poles. Therefore, I submit: the slow moving hard and soft mantles on top of the magma are gravitationally bound to the hard core.

Thus, the magnetized hard core is forced to make changes in its position. Because the gravitational force at the region of the core is so strong, the core resists any change in its position and revolves erratically. When the position of the core shifts, the magnetic poles of the Earth shift.

Core samples taken at Hawaii to determine the Earth's magnetic past (shown on *NOVA* over the PBS network) have revealed that 780,000 years ago, the magnetic poles were reversed. That is, the magnetic force coming out of the Earth was at the south pole and the magnetic force going into the

Earth was at the north pole. The research scientists did an excellent job in discovering the fact that the poles move and shift causing anomalies over the earth. However, no one ventured to satisfactorily explain why the magnetic poles shift, until this work submits the reason.

Consider The Following:

Beneath the surface of the Earth, there are concentrations of iron that are magnetized. When the core shifted, causing a shift of the magnetic poles, these concentrations did not shift. Only a long period of time can rearrange the iron atoms to the new pole arrangement. By the time that occurs, the poles will probably shift again, causing magnetic anomalies to remain.

The combined gravitational force of the Moon, Mars, and Jupiter when conjuncted causes Earthquakes, Volcanoes and causes the core of the Earth to be slowly moved by the force of the moving mantle. In addition, it causes the Earth's rotation to slow, thus lengthening the 24-hour day by about a second at intermittent times,

### GRAVITATIONAL FORCE MOVES MATTER

The center of gravity between the Moon and the Earth is slightly more than 1,000 miles beneath the surface of the Earth, and this point is within the soft mantle. Thus, as the Moon, Mars, and Jupiter become aligned; a great sliding gravitational force is exerted upon the soft mantle since the Moon revolves around the Earth and the Earth rotates and revolves faster than Mars and Jupiter.

There have been motion graphic illustrations given near the twenty-first century that reveals how the water of the seas and oceans are bulged towards the Moon. It is quite impressive to see how the rotating Earth and the revolving Moon creates the tides. Since the Earth rotates with such precision movement as does the Moon, precision tables have been created that reveal the 24-hour 50 minute cycle of the tides. In addition to the moon pulling on the Earth's oceans, the Sun is also pulling and when the Sun, Earth and the Moon are in alignment (new Moon and full moon) scientists have estimated that the Sun is responsible for about 46 percent of the total force that creates the tides.

Although we can see the tides that are created by the gravitational pull of the Moon, we cannot see how the soft mantle and the magma within the Earth are also forced to move according to the gravitational attraction of the Moon, Mars, and Jupiter.

I submit unequivocally: the angular gravitational pull upon the soft mantle and the magma that is caused by the gravity of the Moon, Mars and Jupiter while the Earth is rotating, acts upon the Moho, and the mantle, consequently, the crust is influenced to move.

Although the earthquake forces are uneven forces since only the surface of the Earth that faces the Moon, Mars, and Jupiter experiences the greatest pull. Therefore, as with the tides there are high and low tides, but the magma is compressed by the mantel, crust, and the crust base and cannot freely move with a bulged shape as the seas move due to the Moon's gravitation.

Thus, great gravitational force from the Moon, Mars, and Jupiter causes the soft mantle and the magma to flow intermittently as a very slow moving interior river. Thus, the soft mantel's movement causes the hard mantel to move which causes the crustal plates to move and the core to slowly rotate. The crust, Moho, and the mantle are actually floating on the liquid magma.

Therefore, the Earth's crust, Moho, plus the soft and hard mantle *do* move as earthquakes testify. However, some plates move more rapidly than others, while other plates collide and push up mountains, and at subduction zones large areas, even complete islands, can be forced down within the Earth as plates are pushed over top of other plates.

While earthquakes cannot be prevented, disasters from gigantic volcanic explosions and eruptions *can be* prevented by controlling the forces that cause this natural phenomenon.

Not all volcanoes have a vent or pipe that goes directly to the magma as Kilauea in Hawaii. Some volcanoes release poisonous gases as sulphur dioxide and billows of black smoke. The reason for the black smoke is given here: magma sometimes passes through crude oil, natural gas, coal veins and lignite as it rises through the vents. Therefore, the type of volcanic action is dependent upon the type of earth material the magma passes through on the way to the surface. Water causes an explosion when the magma meets water on its way to the surface. This is the reason for volcanic ash being spewed from a volcano.

Volcanoes are caused by the Earth's gravity force pulling on the Earth's magma by the -- 1. crust - 2. oceans, and 3. the mantle.

The Earth's crust, oceans, Moho, and mantle are continually pulled toward the center of the Earth. This causes great pressure to be exerted upon the magma. Therefore, to release the pressure, pressurized magma is forced to the surface of the Earth through vents. Thus, volcanic explosions and eruptions release the pressurized magma on the Earth's surface as volcanoes.

This pressure upon the magma can be illustrated by the squeezing of an orange. If an orange is squeezed until orange juice flows through the skin, the juice will flow through ruptures at the skin's weakest points. However, if the orange is pricked through the skin to the soft interior by a nail at several places and then squeezed, orange juice will flow from the holes produced by the nails. Of course, in this case, the orange juice flows in a controlled manner. Also, the orange is squeezed from the outside and in the case of the Earth, the crust and the mantle is pulled against the magma from the inside by gravitational force.

## PREVENTING DESTRUCTIVE VOLCANOES

Controlling volcanoes is very simple: release the pressure on the magma. Each year about 15 to 20 of the 60 potentially active volcanoes of the over 1,500 known volcanoes erupt. Therefore, several volcanoes on Earth that do not erupt sulphur dioxide, that is yet to be determined by volcanologists and other scientists as to where and how many places on Earth should be controlled to cause a continual lava flow.

I suggest Montserrat in the Caribbean, Krakatao on a Borneo island, volcanoes on the island of Java, Kilauea in Hawaii and perhaps one in Central America and the Philippines. The six mentioned are excellent places that could be chosen to keep volcanoes continually flowing lava that are away from populated areas and do not erupt sulphur dioxide.

It is necessary to force a constant lava flow. Earth scientists must determine where and how many volcanoes must be kept open to guarantee a continual lava flow. The magma pressure *must* be released by *bombing* an active volcano whenever it ceases its flow of lava. Releasing the pressure on the magma is a very necessary and logical solution, just as releasing the pressure on a pressure cooker to prevent an explosion and in the case of glaucoma, releasing some aqueous humor to reduce excessive pressure within the eye.

As the Earth ages, volcanoes have and will become more

numerous and more powerful because the crust becomes thicker causing greater pressure upon the soft interior. Even in the last century, the number of volcanoes has increased.

Humanity *must* create safety valves to relieve pressure on the magma; the loss of life, personal miseries and the interference of humanity's development has been and will be more devastating to those where volcanoes erupt. When the crushing force from the crust, oceans, Moho, and mantle that are pressing inward upon the magma have temporarily been reduced, as a pressure cooker losing its steam pressure, lava flow will reduce and a volcano will seal the top of the pipe by cooling, but this ceasing of volcanic activity is always only temporary.

The zone beneath the hard mantle moves at a different rate of movement than the crust and this also causes vent tubes to close.

The law of change which I may have originally coined states: Everything changes in relation and proportion to the combined forces acting upon it. Therefore, every second of time, Earth's gravity continues to pull every Earth particle towards its center. Thus, the continual pulling inward of the crust, oceans, Moho and mantle will cause the magma to take the line of least resistance to release the internal pressure and volcanic eruptions will also occur along fault lines. If dormant volcanoes have been sealed by a thick dome, it sets the condition for a violent eruption.

Only by continually releasing the pressure on the magma by inducing lava flows, can violent eruptions be prevented from occurring. Yellowstone, Toba, Laki, Tambora, Mt. St. Helens and in our recorded time period, Krakatao and Vesuvius have erupted with violent eruptions, and could again.

While proving that general relativity is a false theory, energy dissolves, planets revolve in an Aust powered by the Sun are important.

Two projects that should take priority from this work should be: controlling volcanoes and controlling hurricanes.

There are about 1,500 volcanoes in the world. Most of them in Indonesia, and of the total number at the present there are about 15 to 20 that intermittently erupt. However, there are about 60 volcanoes that are considered to be potentially active. Since 80 percent of volcanic action occurs beneath the seas, this control measure would also help to stabilize the temperature of the Pacific Ocean. This would reduce global abnormal weather patterns including excessive rainfall in years when the overheated Pacific waters cause havoc with rainfall patterns. As the crust becomes thicker, it will steadily increase the inward pressure upon the magma and not only does this create the condition for more violent volcanic eruptions, but for excessive El Nino effects, since most of the volcanoes are under the Pacific on the "Pacific Rim of Fire."

Vast highly populated regions in Nevada, Washington, Oregon, and lesser-populated Wyoming, including Yellowstone National Park, could be covered with ash and lava. In addition, Mexico City, plus vast regions in Italy around Mt. Vesuvius are places that could be obliterated by violent volcanic eruptions in the future if positive action is not initiated to reduce a building pressure force upon the magma.

The super volcano at Yellowstone National Park erupted about 600,000 years ago and is now showing signs of eruption development. In the August 2009 issue of the *National Geographic* magazine, Yellowstone National Park was the focus of an article devoted to: When Yellowstone's Super Volcano explodes. Wisdom dictates that this should not be allowed to happen. The loss of life and real estate would be too great to calculate due to ash spewed over hundreds and hundreds of miles plus the drop in the Earth's temperature.

In the case of Montserrat in the Caribbean that started to erupt on July 18, 1995, the pyroclastic flow could have been forced to flow eastward and thereby away from their city of Plymouth. However, not only has the city of Plymouth been destroyed by a covering of twenty feet of ash over the entire city, the pyroclastic flow doomed a large region from ever being inhabited. The volcano has continued to erupt intermittently, the latest eruption as I again edit these manuscripts, occurred at 1:14 AM July 3, 2005, when it sent ash up to 15,000 and 20,000 feet into the atmosphere.

Instead of using controlling forces as careful bombing to cause the volcano to erupt and direct the lava and pyroclastic, if there be any, to flow eastward away from the city of Plymouth, the volcano was allowed to erupt in an uncontrolled destructive manner and caused obliteration to the southern part of the Paradise Island. If the volcano would have been bombed and a path opened for the lava to flow eastward from the volcano before it exploded, the island would have been saved from destruction.

**VOLCANIC ASH ANALYZED**

Scientists have carefully examined volcanic ash and discovered that it is microscopic particles of lava consisting of rock, mineral, and volcanic glass fragments similar to finely crushed window glass. The ash particles are small measuring slightly larger than the size of a pinhead and are extremely abrasive and mildly corrosive. While forcing a volcano to erupt is new thinking as far as I know, controlling the direction of lava flow was performed in 1935 on Mauna Kea in Hawaii. The town of Hilo would have been destroyed by lava since lava was flowing towards the town, but the United States Air Force opened a channel for the lava to flow away from the town of Hilo by bombing the side of the volcano that enabled the lava to flow away from Hilo.

It was a clever maneuver indeed, in 1935, and a blessing for the residents of Hilo. Compare that with England allowing the town of Plymouth to be completely destroyed. Although saving the town of Hilo in 1935, has not been repeated in 2014. Kilauea has been erupting since 1983 and the lave flow has flowed down the mountain to the sea. However in 2014, the path of the flowing lava has changed to flow toward the District of Puna into the Commercial District of Pahoa. Homes and businesses have been evacuated because the lava flow will cover the buildings. I am sorry to say that this is another example of stupidity that has ruled in the Obama Administration and unlike saving Hilo, Pahoa will be allowed to be destroyed.

Yet, as previously given, volcanoes may not be completely controlled, for there is another cause that induces the outward flow of lava. The molten magma within the Earth also takes the line of least resistance as the inward pressure of the Earth's crust, oceans and mantle press on the magma. Consequently, the magma pushes up in chimneys where the crust is thin and it follows fractures in the crust between the tectonic plates and from subduction. The fractures where tectonic plates are together or come together, are the reasons for the volcanic activity in the Pacific called the "Pacific Rim of Fire."

Although, these type volcanoes can be minimized by continually releasing the pressure upon the magma. If controlling measures to release pressure upon the magma had been put into operation prior to 1943, the sudden appearance of a volcano at Taricutin, Mexico on February 20, 1943 on a farmers corn field might not have occurred, and a possible catastrophic land slide of the La Palma volcano on the Canary Island can be avoided.

On April 17-18, 2010, an explosive volcano on Iceland spewed ash over much of northern Europe causing billions of dollars in damage. This catastrophic volcanic eruption could have been minimized and maybe avoided if pressure on the magma had been reduced by forcing a flow of magma from volcanoes where land is uninhabited.

The Four Terrestrials and Our Moon

As we look backward to the first volcanoes, it is a fascinating study. In the very early history of our solar system, all planets were blobs of Sun matter. The first four terrestrials, our Moon and Pluto were the first to cool due to their comparative small size. In our vicinity, Mercury, Venus, Mars, and the Earth's Moon cooled before the Earth.

Therefore, each of these four formed a crust before the Earth formed a crust and volcanic eruptions began as soon as a crust formed. The early surface features of each of these four mentioned was due to their volcanic eruptions. As billions of years passed the interior magma was forced to the surface by the inward pulling of the crust on the magma of these planets and the Moon was reduced in size as they cooled.

The lack of great oceans was also a factor in the forces and the lack of forces that could contribute to reshaping the surface of each. In addition, as volcanic activity ceased, a lack of atmosphere enabled meteors to crash to the surface without being broken to pieces by heat and this contributed to shaping the surface of each.

**OUR EARTH**

In the very early history of the Earth, the Earth was a sphere of Sun matter without any water on its surface. The sun had ejected huge amounts of hydrogen and oxygen with the sun matter and these two elements formed water vapor in the Earth's early atmosphere. There is no doubt that: the Earth was already spinning or rotating on a zero-degree axis just as a fast-moving top spins.

As soon as the Earth cooled enough to enable the water vapor that was in the atmosphere of mostly nitrogen gas to condense into water, it began to rain and it rained continually for thousands of years. During the first phase, the rain quickly turned to steam and returned to the atmosphere. During the second phase, some very hot water began to collect in large pools and any crust beneath the water became thicker.

Eventually, during the fourth phase, the Earth was completely covered with water over a thin crust. As soon as the water covered the Earth, the gravity force pulling the water and the thin crust towards the center of the Earth pushed upon the molten magma just below the water and the thin crust, and volcanoes began. This continual volcanic eruption for several billion years thickened the outer crust and thereby created the first land mass. Slowly, as the crust thickened a secondary crust or mantle formed. As the land mass grew, pressure upon the magma increased and volcanic activity increased. As several billion years passed, new volcanoes erupted on top of old volcanoes, the land mass grew or expanded and its crust became thicker.

Centrifugal Force

After the crust became a mass that exerted an additional force to the water upon the spinning Earth, an imbalance in the rotary force created another force that worked upon the land mass to achieve rotary balance or equilibrium. Therefore, two situations began to occur. First: the land mass that had grown to become one large continent had been exerting a counter force upon the centrifugal force of the spinning Earth. This can be illustrated by the following analogy. If you spin a top after you place a very heavy glob of grease on one spot, it will create an imbalanced force to the centrifugal force and the axis will be altered by the imbalanced force. In addition, the glob will separate and move on the spinning top to cause a rotary balance or equilibrium.

Secondly: (according to my understanding) in addition to centrifugal force, the gravitational pull of the Moon, Mars and Jupiter broke the crust into huge plates to achieve a rotary balance or equilibrium.

**THE SPINNING EARTH TILTS 23.5 DEGREES ON ITS AXIS**
Once equilibrium, in the centrifugal force was being reached, according to the distribution of the land mass, the tilting of the Earth on its axis ceased at 23.5 degrees. What a strange world we would live in if the Earth did not tilt on its axis, for the Earth's tilt is the reason why we have changing seasons. It is logical to attribute the uneven distribution of the land mass as the reason for the Earth's 23.5 degree tilt. The land mass of the Earth being unevenly distributed not only causes the out of balance spinning condition, I suspect this is the reason for the Earth's wobble. However, earthquakes are causing tectonic movements towards a state of equilibrium within its centrifugal force. The distribution of the Earth's land mass is changing and over millions of years the Earth's 23.5 degree tilt will change as the changing equilibrium changes the tilt. If volcanoes would

erupt at the South Pole or an excessive amount of ice would form at the South Pole, it would change the 23.5 degree of tilt of the Earth. All this given data about land mass movement, volcanoes, and incorporating the known 23.5 degree tilt of the Earth is my concept, and I consider it far superior to any other presently held concept.

The uneven cooling of the Earth also had a very minor influence in the breaking of the plates and plate movements. In regard to naming these super continents, I respect the name of Pangea or Pangaea that was submitted by the German geophysicist, Afred Wegener in 1912; however, the suggested time for the first super continent to break into separate land masses being 280 to 225 million years ago seems doubtful to me.

This time period is questionable since thus far no data has verified this. However, I do not accept the idea that several huge continents joined after separation. The name Gondwanaland has been given to this theoretical huge continent. *If* two huge continents did collide and become one, the tectonic forces (centrifugal and gravitational forces) would exert a force to bring a new balance to the Earth's spin by land mass movement.

**NASA DATA**

The NASA website, volcano.und.nodak.edu has extensive information about the history of volcanoes and present-day active volcanoes. You will not find anything about how to control devastating volcanoes at that site, but rejoice, you have found that information in this work.

On August 29, 2010, Sumatra's Mount Sinabung erupted after 400 years of dormancy. Pictures of people covered with volcanic ash circulated over the Internet. This volcanic eruption could have been avoided if selected places where controlled volcanoes had been caused to release the pressure on the magma. Indonesia's Mount Merapi erupted in the last of October 31, 2010, and into November. People were killed and ash spread for miles away. The devastation was referred to as the Pompeii of Indonesia. It too could have been avoided if selected places had released the pressure on the Magma from the Earth's crust and mantle pushing inward on the Earth's interior. The same condition could occur at Yellow Stone National Park if effort is not made to reduce the pressure on the magma AND thereby control volcanoes.

## ICE AGES

Secular scientists are now turning their research to the gravitational force of the Moon as a source for 'Ice Ages.'

In the August 15, 2000 issue of *Science News*, an article titled "It's High Tide for Ice Age Climate Change" reveals several statistics about the Earth/Moon relationship that can cause "high tides," and therefore bring much cooler water from the depths of the ocean to the surface. The article states that tides are exceptionally high every 411 days when the Moon comes closer to the Earth or is at perigee.

Also, the article states that every 1,800 years when the Earth is at perihelion (closer to the Sun) while the Moon is closer to the Earth (perigee), during New Moon, excessive high tides will result and this could result in a climate change, although the position of the Earth and the Moon as given will cause high tides. However, I do not accept the probability of the Moon's gravitational force being the key element in climate changes for the combined force of the Moon, Mars and Jupiter when aligned in conjunction has far greater power to cause very High Tides, Earthquakes and Volcanoes.

Also, consider: the greatest factor in the warming climate change is the green house gases created by man's air pollution. However, an opposite condition can occur from ash in the atmosphere. This can be caused by erupting volcanoes and was indeed the prime factor in ages gone by in preventing the heat rays from the Sun to warm the Earth.

Thus, the Earth's climate was altered to become *cooler*. On the basis of this known affect, I consider a sole Earth, Moon relationship to have minimal effect upon the Earth's climate. However, please keep in mind; politicians in America and China are responsible for contributing to the Earth warming. The surface of the Earth is warming from greenhouse gases, chiefly from the billowing smoke from volcanic eruptions, peat fires in Indonesia spewing millions of tons of greenhouse gases, the smoke from burning rain forests, and smoke from coal fired electric generation plants burning of fossil fuels and the exhaust from automobiles and trucks.

February 24, 2014. I just received an email letter from my cousin–in–law in Utah about "Earth Warming," and I was so enthralled about its contents that I am inserting the gist of the letter. This scientist's comprehensive understanding about "Earth Warming" dovetails with my understanding of the carbon dioxide spewed into the atmosphere by volcanoes and how to control volcanoes so nicely that I am including his remarkable revelation in my manuscripts.

This Australian scientist's name is Ian Rutherford Plimer. His entrance into our domain was on February 12, 1946. Mr. Plimer became interested in geology and devoted his life to studying geology. Not one to sit on his you-know-what and loaf, he taught in universities and wrote extensively about different aspects of geology. The University of Melbourne is honored to have Dr. Plimer as a professor emeritus of earth sciences at the University of Melbourne.

In addition, he is a professor of mining geology at the University of Adelaide, and the director of multiple mineral exploration and mining companies. He is a prolific writer having published 130 scientific papers, six books and edited the *Encyclopedia of Geology*. He has revealed some very important facts about Earth Warming. He stated: "The volcanic ash emitted into the Earth's atmosphere in just four days - yes, FOUR DAYS - by that volcano in Iceland has totally erased every single effort you have made to reduce the evil beast, carbon. And there are around 200 active volcanoes on the planet spewing out this crud at any given time, EVERY DAY.

I don't really want to rain on your parade too much, but I should mention that when the volcano Mt. Pinatubo erupted in the Philippines in 1991, it spewed out more greenhouse gases into the atmosphere than the entire human race had emitted in all its years on Earth.

Yes, folks, Mt. Pinatubo was active for over one year - think about it!

Therefore readers, I already have one prestigious scientist in agreement with me or I agree with him, it is up to politicians to use our resources to control volcanoes.

Switching To Hydrogen as a Fuel Will Cause Great Beneficial Changes to the Earth's People and to the Earth.

If my great gift to the American people will be received by politicians and locked by legislation, our environment and the wealth of the law-abiding nations, will be greatly improved.

**OCTOBER 29, 2012**

A new event has deeply troubled my mind: the destruction and misery to thousands of people brought about by Hurricane Sandy has brought power from on high to reveal how to control hurricanes.

Read the thoughts that come to me from on high.
***Cities can be saved from destruction.***

**CONTROLLING HURRICANES**

Controlling hurricanes can be done by artificially creating conditions normally created by nature.

**THE NATURE OF A HURRICANE**

Coriolis's work was incorrect when it credited the Earth's rotation for causing counter clockwise motion of atmosphere and liquids (going down a drain) in the northern latitudes and clock wise motions in the southern latitudes. Coriolis did not understand the nature of an Aust neither do scientists that followed him up to present day. Therefore, they were only partially correct when they attributed the physical motion of the Earth to be the sole motivating force that produced both these clockwise and counterclockwise motions of atmosphere and causes water to swirl as it goes down a drain.

If The Earth Did Not Rotate, Water Would Not Swirl As It Goes Down A Drain. Please Understand: It Is Gravity That Pulls Water Down The Drain And Nothing Else.

Since you the reader are already knowledgeable about the Sun's Aust, it will be easy to understand how hurricanes are formed. Remember: an Aust is an extension of rotary gravitational forces radiated from the central girth's greatest diameter of a rotating celestial body where the central girth is 90 degrees to its axis. The Earth is continually radiating gravity waves to outer space. However, the Earth is rotating and that reformulates the gravity waves to conform to an altered spiraling or screw like pattern.

The most powerful and the fastest -moving gravity force will create a commanding region in the radiating gravity waves. This most powerful region is an extension of the gravity force at the Earth's greatest girth and that is at the equator. The rate of rotational movement or rotational speed at the Earth's equator is about a thousand miles per hour. Directly above and below the equator the rate of movement becomes slower since the circumferential distance is less than at the equator. As lines are traced from the equator to the north and south poles, the circumference diminishes. Therefore, at each line of latitude north and south of the equator, the land mass travels less distance as the Earth rotates. Again—since the land mass above and below the equator travels less distance as the Earth rotates, the rotating motion of the Earth is the same (once every 24 hours) for all latitudes. Therefore the decreasing latitudes moving less distance in the same time is stated in a formula—DxT= rate of movement. Distance multiplied by time equals the rate of movement; therefore each latitude away from the equator is moving slower.

An Air France flight 447 from Rio de Janeiro to Paris disappeared over the Atlantic on Monday June 1, 2009 with the loss of 228 passengers and nine crew members. The area where the plane disappeared is known in meteorological science as the Intertropical Convergence Zone. It is a zone where different air streams converge or cross as the atmosphere moves due to the rotating Earth. It is not coincidental to also discover that Flight 447 was at 1.5 to 1.7 degrees N. Latitude. Remember—it is not the Coriolis affect that causes water to swirl in counterclockwise directions and clockwise directions. *Gravity* causes water to swirl as it is pulled down a drain, not the Coriolis affect.

Flight 447 was within the Intertropical Convergence Zone *and in the Aust* when it became unstable and crashed. The black boxes reveal that up and down motion of the plane caused an icing of the speed indicators and an inexperienced pilot was responsible for the quick steep descent. However, the knowledge concerning gravity in the Aust and above and below the Aust opens new factors to be considered by meteorological scientists.

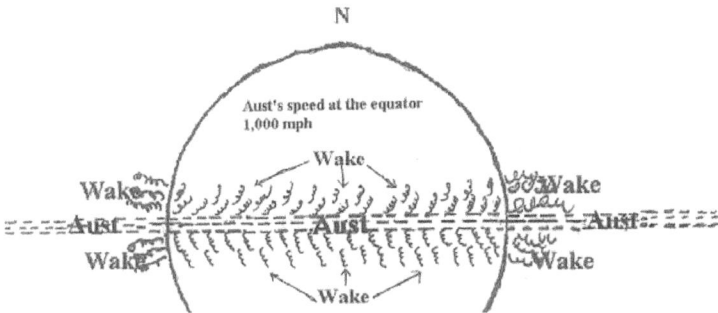

The radiated gravity forces at the zero line of latitude or on the equator interact with gravity forces above and below the zero line of latitude. Therefore the distinctly different gravity forces interact with each other as two streams of water traveling at different rates of movement contact each other. When this happens, strong eddy currents are created at the region where the two streams make contact. This is exactly what happens from the Aust force or fastest moving gravity waves meet and interact with slower moving gravity waves. This was explained in detail when the Aust was explained earlier. Therefore, in the northern latitude just above the boundary where the Aust force meets a slower moving gravity force, a counter clock wise gravity force is created, or gravitational eddy currents. Also, another very important contributing factor to the formation of hurricanes is the temperature of the water (already given.) This is why the dynamics of a hurricane is similar to a—dust devil.

The warmer the water, the more rapid the water molecules are moving about in the body of water causing more molecules of water to escape into the atmosphere. However, the rising vapor is in a gravity field on Earth and at that region above the equator the gravity waves are spiraling in a counter clockwise direction and a counter clockwise rotation of warm moist atmosphere will be initiated. The opposite will occur south of the equator.

The gravity force of the Earth will surely not subside and the eddy currents in the Wake above and below the Aust will not subside; therefore once a hurricane forms a central hub or eye, the warm waters will feed the - rotating vapor eddy - or hurricane.

Only when the hurricane moves north, farther away from the powerful eddy current forces and is not fed by warm tropical waters, will the hurricane lose its rotating motion and become a tropical storm, and eventually die.

And so scholars of science I repeat, all the present books about meteorology are now obsolete.


**CONTROLLING HURRICANES**
Controlling hurricanes can be done by creating conditions normally created by nature.

## THE NATURE OF A HURRICANE

Hurricanes are huge storms that have four prevalent conditions.

One—in northern latitudes, they swirl in counter clockwise movement

Two—they are heavy laden with moisture - (water).

Three—they are composed of warm air

Four—the atmospheric pressure is lower than the average sea level pressure because warm air is more rarefied and thus the air molecules being more widely separated exert less pressure than cooler air where the molecules are more tightly packed.

Thus, higher pressure and cooler air is able to affect lower pressure warmer air.

Historical records reveal a hurricane that was approaching the gulf coast states was deflected by a high pressure area that was moving or was stationary at some distance inward from the coastal region. In addition, the hurricane named Sandy that struck the East Coast on October 29–30 2012 that caused schools to close, mass evacuations of 375,000 people from flood-prone neighborhoods in Brooklyn and Manhattan, caused businesses to close, brought power outages to seven million homes and businesses and in Connecticut brought the worst flooding in 70 years and 122 deaths; all hurricanes could have been prevented The total cost from the damages from Sandy that can be measured in dollars is estimated to reach 50 billion dollars: the emotional damages to families is so colossal it is inestimable.

The course of the hurricane named Sandy was altered by two high pressure cold fronts, one west of hurricane Sandy and one north. These two cold fronts deflected the warm air low pressure storm to the heavily populated coast and this caused the hurricane to remain over the heavy populated area and this intensified the hurricanes destruction.

**THE BIRTH OF A HURRICANE**

Hurricanes and typhoons are natural creations from waters whose temperatures are 80 degrees and more in the outer fringes of the earth's Aust. Every rotating sphere has a complex gravity field. At its zero line of latitude it has an altered gravity field that is polarized in a straight uniform field and hurricanes cannot form in this region. Above and below the polarized gravity field, the fastest moving polarized gravity field interacts with slower moving gravity fields. This interaction of fast and slow moving gravity field creates counter clockwise and clockwise gravity swirls that I have named the Wake Zone.

* When an area of warm, moist water becomes a low pressure area it lifts slightly from the water's surface. This is the critical phase in the birth of a hurricane - because - when an area of warm moist water lifts from the surface of the sea or ocean, it lifts within the - Wake Zone. Gravity in the Wake Zone in northern latitudes is swirling in a counter-clockwise direction.

*When an area of low pressure lifts - ambient air pours into this area and the low pressure area is stimulated to begin to rotate in the counter clockwise direction. The swirling gravity force of the Wake is the driving force that creates a hurricane. If the Wake was nonexistent as at the equator region -- the storm would be a tropical rain storm and would not swirl.

A free falling body will accelerate until it reaches its 200-mile per hour terminal velocity. Therefore taking that message from nature it is evident that a swirling gravity force can exert about a 200-mile an hour rotational force on a hurricane. Although atmospheric currents have a complex nature and there are forces and counter forces working upon hurricanes that 200-mile per hour wind gusts are rare.

Once a hurricane is formed it does not stay stationary; it moves. Hurricanes do not just lay on the warm waters of the Caribbean or the Gulf of Mexico as a dust devil does on the western plains, the height of hurricanes reaches high above the atmosphere where it joins an existing air current or is captured by an air current. These very high air currents are the steering mechanism for hurricanes.

The atmosphere has many air currents that affect weather on the Earth and they are in continual motion. Two of the best known are the Jet Stream and the Trade Winds. The Trade Winds flow from east to west and are the dominate air flow that guides or pushes developing hurricanes from Africa to North America and to a lesser extent to South America. African hurricanes are born in the Sahara desert and off the coast of Africa in the general vicinity of the Cape Verde Islands. This is about 15 degrees above the Equator and in the region where the gravitational Wake exists. Once an African born hurricane reaches the Caribbean it gains ferocity by the warm waters of the Caribbean and their movement is influenced by any cold fronts and the high spiraling air currents in the Caribbean named the Bermuda High. Although the African born hurricanes are the most destructive, they are not the most numerous. Most hurricanes are born in the Caribbean and their steering mechanism or forces that steer them is the same as the African born hurricanes.

The Bermuda High is responsible for steering hurricanes away from Florida and the southern coast and steering them up the eastern coast to the New England area. The steering air current of the Bermuda High is not a fast moving air currents and *hurricanes are very sensitive to any steering forces.* Infrequently there is a high-pressure storm or cold front that is located off the coast at the interior of land as occurred with the hurricane named Sandy. Hurricanes cannot penetrate a high-pressure cold front any more than light can penetrate a cement wall. As shown by historic events, hurricanes will be deflected by a high-pressure cold front. And I repeat: hurricanes cannot penetrate - a—high pressure cold front any more that a person can ram their fist through a brick or cement wall. *A high-pressure cold front will deflect a hurricane.*

This High Pressure Cold Front Condition Can Be Created Artificially And Will Have The Power To Deflect Hurricanes.

Here is how manmade blasts of cool air swirling in counterclockwise direction or clockwise which ever proves to be the best, travelling at about 150 to 200 MPH pointed towards a hurricane will deflect a hurricane.

### HOW TO—FIRST IDEA

Giant wind-tunnel type fans fifty feet in diameter with three or four blades belt driven by a powerful internal combustion truck engine and pointed to the hurricane with a bank of three to five sections of a bank of *giant refrigeration coils* 45 ft. to 55 ft. in diameter - and - if needed - with snow producing machines as used by Hollywood and Ski Slopes with the nozzles on front of the fans. The snow would melt immediately as it is swept with the blast of air although it should decrease the temperature to less that 40 degrees below the ambient atmosphere -- And This Will Deflect A Hurricane.

Test And Proof Device: at first five or six flattop ships with ten or more fans separated by about 30 to 50 feet from each other and pointed upward at one degree and thirty degrees toward the approaching hurricane. The test machines can be made without the snow producing snow, this should prove that hurricanes can be deflected. After the idea is proven, ten hurricane deflector ships can be built. Some fans can be pointed across the horizon from three o'clock to nine o'clock on the compass.

There is no maybe -- special flattop ships equipped to be hurricane deflectors should be built.

If a hurricane continues north and as it moves it would graze the Eastern Coast as it moves, five or six hurricane deflector ships equipped with the fans (air going 150 to 200 miles an hour in counter clockwise direction or clockwise which ever would be the most effective.) Cowling surrounding three Johnson Controls Superchillers for each fan to chill the air (to 40 degrees or more below the temperature of the ambient air) that is pointed toward the hurricane would deflect the hurricane. If the hurricane continues moving north, the hurricane deflector ships could move north and continue to deflect the hurricane away from land. Ships move faster that the forward motion of a hurricane so a hurricane could be continually deflected away from land.

Land-based deflectors would not be the most effective nor the most desirable. Made man islands away from the shore is the most favored design. Flattop ships are the most desirable mobile method; however they would have to stay close to land to keep any outer bands of the hurricane from making land contact. One fan on the ship would have to point SSW to keep outer bands from making contact with land.

Although a super storm as Sandy that is 900 miles long may not be able to be completely deflected; that remains to be seen. It would have to be deflected when far out from land. For sure densely populated areas could be protected, although this does not downplay deflecting hurricanes for a 900-mile long hurricane with a diameter of over 300 miles makes a monster hurricane an anomaly. Saving billions and billions of dollars in damages by most hurricanes is a super blessing for humanity.

Although the old adage "an ounce of prevention is worth a pound of cure" is true for controlling hurricanes. I recommend specially built hurricane deflecting ships should go to a distant hurricane in the Caribbean and point their chilled counter clockwise or clockwise blast of cold air, whichever is the most effective, to deflect the hurricane. The ships will have to be specially designed since the blast of air will push the ship backwards. Therefore front pods would pull the ship forward and pods in the rear would push the ship forward

In the future, in order to protect Florida, Texas, Mississippi, Louisiana, and Alabama, several half-mile-long artificial islands separated by several miles with permanent fans and chillers would be ready to deflect a hurricane when one threatens Florida and the Gulf Coast states.

I predict the ferocity of hurricanes can be not only greatly reduced but can be—deflected outward to the Atlantic Ocean and the Gulf of Mexico. Hurricanes are very sensitive to high pressure with cold air as naturally occurring cold fronts have proven.

### HOW TO -- MY SECOND IDEA FOLLOWS

My second idea modifies the cowling around the chiller. In both cases the cowling should be at least as wide as the diameter of the fan blades. My second idea includes a screw type or worm gear type shape on the interior of the cowling, this would swirl the blown air as it is blown through the cowling. In both cases, at least three powerful chillers in back of each fan would be used and if needed, a snow-producing device would be used in front of the cowling where the blown swirling air exits the tube. This should reduce the temperature of the blown air to be 40 degrees less that the ambient air. In both cases, each fan is to be belt driven by a powerful truck motor. This second idea is superior to the first phase deflector for it allows a greater control to point the "hurricane deflecting" air in a desired angle. In addition, the swirling air would give a greater punch to the hurricane.

Readers: most of the members of humanity have imaginations or dreams as to what life would be if our dreams were realized. Well - scientists also have dreams. I have two dreams about hurricane deflector ships. First: I dream of about ten flat-top ships designed for going close to the eye of a hurricane. Huge fans should be located on the periphery and the ship to be driven close to the eye of the hurricane. The huge chillers and snow should drop the temperature from the ambient air to 40 degrees or less and that should fragment a hurricane and ten ships should destroy a hurricane and reduce it to a tropical storm.

My second wish is for a larger deflector ship to be built named the TRUET with a statue of CHRIST on the leading edge. This idea is just and appropriate because after I fasted for 24 days, I was blessed with an extra or psychic vision. I see a portion of reality that the average person does not see and at intermittent times I hear the thoughts of people and know what they are thinking. I also see flashes of - super natural light - that I have learned to know that it is - the Holy Spirit. As I learned to increase my communication to the Lord - the flashes of light increased - but - also at times when I was asking the Lord a question - my screen of vision became dark. So - I can state with 100% honesty - the Light has guided me in all my research. Therefore - I state with 100% honesty, the idea to build hurricane deflector ships came from Christ the Lord. Thus - in consideration of the fact that the greatest revelations in the last two thousand years have come through me from Christ the Lord and God the Father- why not believe me and have a statue of Christ the Lord on the largest and lead ship?

**ALSO CONSIDER**

All of America's war ships are named after presidents and military people for the purpose of wreaking destruction to nations that a president wants to hurt badly. Spain, Austrian-Hungarian Empire, Vietnam and Iraq were destroyed and great suffering was inflicted to these nations to the whims of presidents. A huge hurricane ship, if named the *Truett*, it would prevent destruction and enable a safe and peaceful life to millions of people,

And that readers, is the method to control hurricanes by deflecting them away from land. But it requires the government to put the plan into operation. In review—what will they do? I suspect that I will hear the same type bull s____ as I've heard for decades. This so-called self-styled scientist who claims that Einstein's theory is false and astrology is true will now be told his hurricane proposal is preposterous. It can be compared to a mouse blowing its breath against an elephant. This time readers -- demand to know the names of the doubters.

**CONCLUSION AFTER AN INSERT ABOUT HURRICANE HARVEY AND IRMA.**

But first let us conduct a question and answer session as though we are in a class room. The students will ask questions and make any comment at the close and I will answer the questions.

Question 1: What is a hurricane? Answer: a hurricane grows from a tropical depression rain storm that swirls with winds revolving around its center from slightly over 39 mph. When the swirling winds reach 74 mph it is classified as a hurricane. The swirling winds can swirl at a hundred miles an hour to over 175 miles an hour and swirls in counter clock direction above the equator.

Question 2: What causes it to swirl in counterclock direction?

Answer: Gravity causes the hurricane to swirl in counterclock direction. Remember: water swirls in a counter clock direction as it goes down a drain in northern latitudes - that is -over seven degrees above the equator. In both cases it is nature or gravity that causes the counterclock swirling motion of a hurricane and the counter clock swirling motion of water as it goes down a drain in northern latitudes and clock wise swirling motion in southern latitudes.

Remember on the equator, to about seven degrees on both sides of the equator, hurricanes cannot form and water does not swirl at all as it goes down a drain. The water goes straight down a drain without swirling. South of the equator a hurricane swirls in - clock wise - direction and water goes down a drain in a clockwise direction

Question 3: Since nature or gravity causes a hurricane and water going down a drain to swirl in counter clock direction, in northern latitudes and to swirl in clockwise direction in southern latitudes, what force is causing the swirling motion?

Answer: As stated previously, in both cases, water swirling as it goes down a drain - and a hurricane - swirling - is caused by the force of gravity. However, the force of gravity is not uniform at every place on the Earth.

Question 4: Will you please explain that?

Answer: The reason for the non-uniformity of gravity is revealed in the dynamics of gravity. That is, there are several forces that affect the force of gravity. Let me explain: every particle exerts a gravitational force, although on a celestial body, every particle is a collective part of the greater body. So the mass of the Earth is radiating a gravity force outward from the center of the Earth to every point on the surface of the Earth and outward into space.

However, the two vital factors that causes the Earth's gravity to be non-uniform - are the facts that the Earth is a sphere - not as perfect sphere for the Earth is bulged at the equator region and the Earth is rotating. These two factors are the prime factors that govern the dynamics of the Earth's gravity. I'll explain further. When a flow of air or water pass each going at different rates of movement or speed, a vortex is formed. When a fast moving stream of water passes a huge rock in the stream, the water slows as it goes around the rock and as it joins water on the downstream side the slowed water joins the faster moving water and an eddy current is formed or a vortex and that is a whirlpool. So take notice here: when a flow of fast-moving water joins a flow of slow -moving water, a vortex is formed.

The same condition exists with fast-moving air currents meet slower moving air currents thousands of miles above the Earth. When this happens a vortex is formed and as we see it on the surface of the Earth, it has a funnel shape. When this vortex extends to the surface of the Earth; it is a - tornado. These examples are given to show what happens when a fast moving medium passes a slow moving medium, because—and take special note here—when a faster-moving gravity force meets a slower-moving gravity force - a gravitational disturbance causes a gravitational swirling in a band around the Earth. I have named the gravitational disturbance, the Wake. where the fast-moving gravity force meets a slower-moving gravity force

The question given: What causes the swirling of a hurricane - is not easily answered. I have given you a portion of the reasons why hurricanes swirl, but to present a better understanding why hurricanes swirl, it is necessary to better understand the dynamics of Earth's gravity force. I did state that gravity is extended outward from every point on the Earth. I also stated that the two prime factors that govern the non-uniformity of the Earth's gravity are the facts 1. that the Earth is a sphere although slightly bulged at the center of the sphere and 2. the Earth rotates.

We could create a model of the Earth to show how the Wake is formed - because - a model and a picture is worth a thousand words. I will draw a near picture of the Earth on the blackboard with its Aust and Wake.

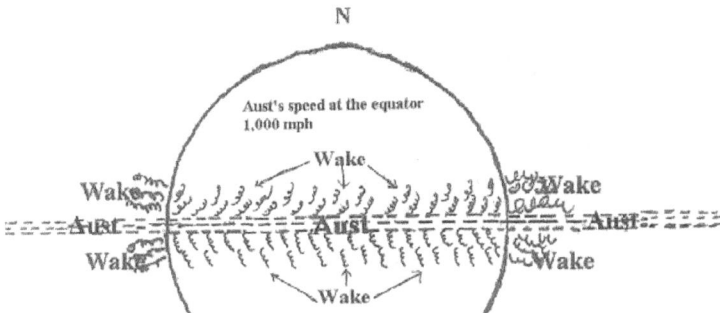

In round figures, the diameter of the Earth at the equator is not quite 8,000 miles and is slightly more than 26 miles more than the distance measured at about ten degrees latitude above its zero line of latitude. As the Earth rotates the surface at the

equator is moving at about a 1,000 mph. However—and take notice—as a line is traced from the zero line of latitude to the North Pole, the distance around the Earth at each latitude reduces. This reveals a diminishing rate of movement or speed since a distance traveled divided by its time equals its speed.

Now we have the rough measurement to prove why hurricanes swirl because hurricanes are formed above the Aust within the region of the Earth's Wake.

We could build a hollow wooden ball model to actually see how the Wake is formed by having holes drilled equally spaced around the ball to allow controlled compressed air to escape from the holes. The spinning motion should be able to be controlled too because if the ball spins too fast the exiting air will not be able to interact with the air that exits above and below its zero line of latitude and a dark pink colored air should be used and piped into the ball from its south pole to see the interaction of fast moving air and slow moving air.

The explanations given should suffice to answer the question, why do hurricanes swirl?

Thank you. Ques: can hurricanes be destroyed or controlled? Ans: No, Hurricanes cannot be destroyed or controlled; however by using the lessons from nature, hurricanes can be steered by creating a controlled high pressure force that can deflect hurricanes away from land.

I state to all that are interested: the 180 billion dollar damage to the state of Texas could have been avoided by using about ten flat-top ships equipped with hurricane deflecting equipment to steer the monster hurricane named Harvey away from land and allowed to become exhausted in the Caribbean.

I state unequivocally, without a doubt all the pain, misery and destruction could have been avoided if the hurricane-deflector ships as I described would have been in place and deflected the hurricane away from land - The Destruction Could Have Been Avoided.

Perhaps this book will result in ending the terrible destruction hurricanes wreak upon our great nation. The ships will be expensive to build but they will cost only a fraction of the damage wrought by Sandy, Harvey, and Irma. Hurricane ships will be one of the greatest blessings ever bestowed upon humanity and readers - thank the Lord in your prayers.

**CONCLUSION**

Scholars of science of the twenty-first century and beyond: for those that have read this entire volume titled *New Physics - New Astronomy*, I congratulate each of you and I have asked our heavenly Father to bless your minds with an expansion that can only come from our creator and his only son Lord Jesus the Christ. I hope your intellectual journey into the remaining volumes is just as rewarding or more than the reward that you experienced while reading this volume. I urge each of you to stay with this intellectual journey. Keep your thinking cap in place and remember, what you put into your mind while on this earth is directly related to what you take out of the earth when you depart these physical environs.

Each of you is now enabled to take a part in building a new science. Use your legal influence to demand universities to make the tests that are explained in this volume. When they do make the tests and the results are published in disclosures to the world's most reputable science organizations, a new science will be born, this will stagger the members of a very staid organization.

Then, when the fact that astrology is also verified by the controllers that disseminate knowledge, peoples of the world will be able to rejoice, for knowing truth is freedom from ignorance.

## CLOSE

Take a deep breath, readers. For the first time in human history, mankind will be set free from the devastating effects of volcanoes and hurricanes and accessing hydrogen as the basic energy source will bring the greatest boost to the standard of living than ever before experienced, for all civilized nations. Also consider, after about 10,000 years of man's history while looking at those lights, in the sky, and wondering if they have an effect on human life, astrology is proven to be true; this proof is revealed in Volume II. It is one of the great times in the history of man to rejoice.

The nature of the electron has been revealed, the law of conservation of energy has been proven that it is not a law but is an unproven assumption. Einstein's theory of general relativity has been proven to be a false theory. Volume I proves that energy is destroyed, or as I state, is dissolved. But please remember that energy is a quality and not something physical as a rock or a log. The law (or opinion) that proposes: "Energy cannot be created or destroyed, known as the "conservation of energy" will stand until someone figures out how to create matter from nothing, or from ether, and on that I am working.

An Atomic bomb destroys some matter therefore remaining is the fact that physical matter cannot be created. Although generating electricity creates electrons from electromotive force, and electrons have an infinitesimal amount of mass and that is creating energy OR creating substance from tenuous electromotive force. That fact should jar every staid physicist, because the ideas embodied in the "conservation of energy postulate" are nincompoopism.

This work with one swath will make every general science book, every physics, astronomy, astrology, meteorology and psychology book plus all encyclopedias obsolete. Volume III will jar all those working in anthropology and maybe theology. I entreat all to open their minds to truth for truth will set you free from entrenched false ideas.

Volume IV reveals the changes that must be made to our financial, political and social structure to Create A Great Civilization. I implore every intelligent and serious American to buy Volume IV titled *Revised Deep Truth* proof is presented that reveals Islam is not a religion; Islam is a philosophy.

The Prophecies Of The Mayan Great Age Ending Is Coming True.

Let us join in a toast: we can use fruit juice for the toast. In addition, the cultural and political changes that will follow within a decade after this work is published will be recorded as the end of an era and the beginning of a new era. However, keep in mind that it will be people power that will cause the changes. Bring the rulers into tow; the power of the people can do it.

Volume III Will Explain Who You Are, What You Are, Why You Behave As You Do, Where You Came From, And Where You Are Going, although before reading Volume III, you should understand how planetary interactions affect human behavior. After grasping the principles given in this volume which you are now reading, you should understand how and why the Aust is created and how it operates within our solar system. The Aust enables the orbs to be created. But first, a 2,500-year historical review about solar system knowledge will broaden and deepen your thoughts about the struggles encountered by scientists in their efforts to serve God and Christ as a revelator about the nature of the solar system and the environment.

Come; let us go to Volume II, *Music of the Spheres*, for your intellectual journey will be a most memorable experience of your lifetime.

www.ingramcontent.com/pod-product-compliance
Lightning Source LLC
Chambersburg PA
CBHW022053210326
41519CB00054B/324